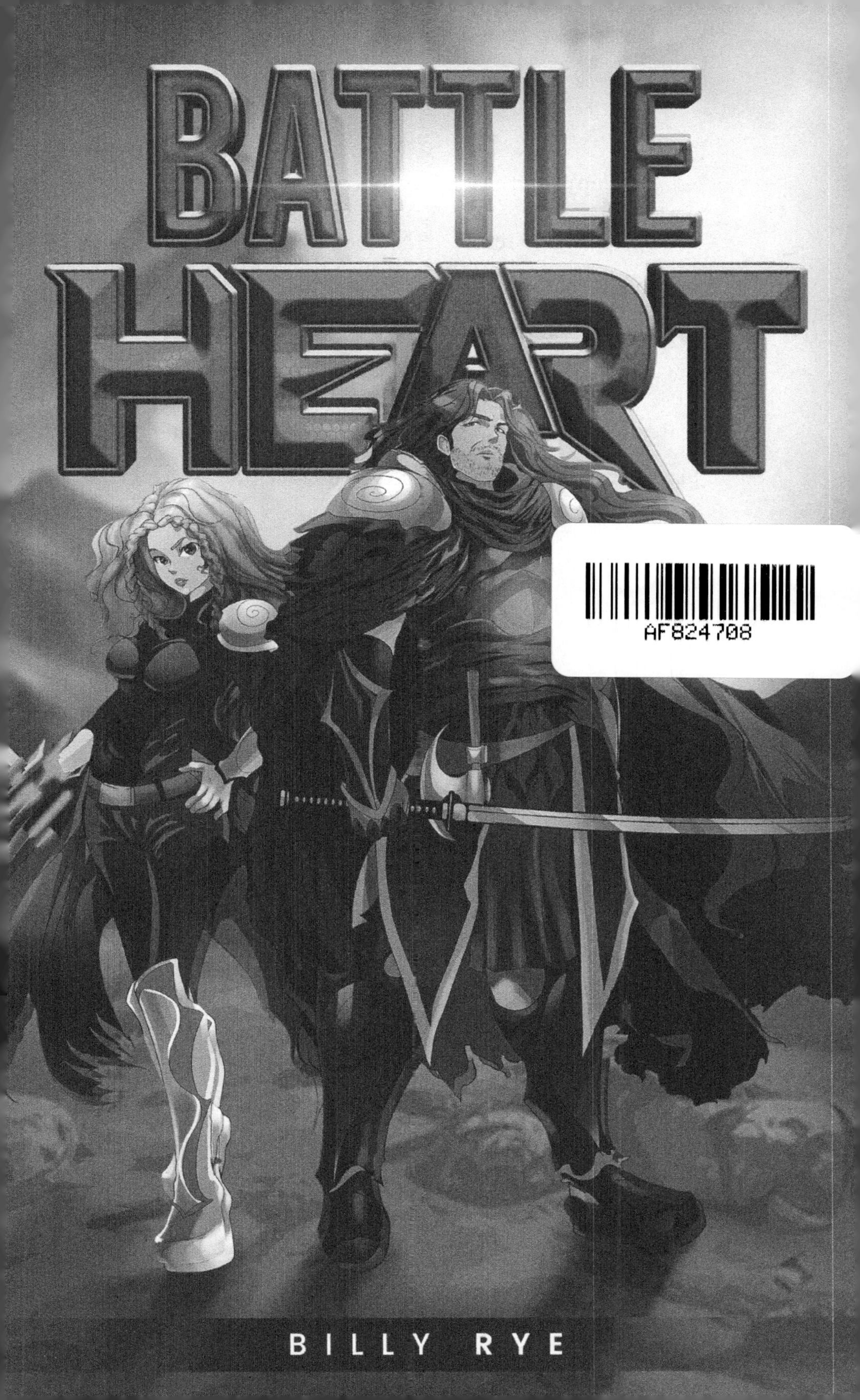
BATTLE
HEART
BILLY RYE

Any references to historical events, real people, or real places are used fictitiously. Names, characters, and places are products of the author's imagination.

To Casey, my battle bride and the joy of my heart.

To my kids who keep my heart strong, Deborah, Megan, Bailey, Lyric, Ashton, Mick, and Cerra.

To my teachers who taught me to perceive and grasp things more, Mrs. R Shulze and Ms. Pickens.

To Ms. Kim and Ms. Linda.

Contents

Chapter 1:
The Partners

The sun slowly set behind the mountains, casting light rays on the city of Italica. Inside the café, a woman sat beside the full-length window, sipping coffee. Her chocolate brown eyes twinkled in the sunlight, giving them a soft but sharp look; her vision was unfocused but full of determination. With every breeze, her wavy black hair flowed in the air as she glanced outside the window. The woman, Cass, was wearing a white ruffled shirt with black pants and looked to be in her late twenties.

Beside her sat an older man in his late fifties. The black hues of his eyes were cold and distant, as if hiding many secrets behind them. The man, Tor, was a foot taller and heavier than most men. His dark—with a hint of white—hair was well-kept. And the ruggedly handsome jawline gave him a smoldering look.

Cass looked toward the sunset as the last rays pierced through the city streets. "Hope and Despair café" was scratched on a simple signboard that rattled with the breeze. It was a great place on the outskirts of town.

What an odd name. Cass thought and glanced over at her partner Tor. *He is not like other people I have met in my twenty-six years.* She watched as the man leaned back in his chair.

Tor and Cass had worked for the Broken Heart Service company for two years. The partners' job was simple when a contractor couldn't pay back the debt, they recovered it by retrieving the body part the contractor had loaned. During those two years, Cass and Tor made a lot of recoveries together and were the best at it. It was a good business for the service company. Cass didn't know all the setup but knew that she and Tor made up a good team.

Tor was not just an ordinary man; he was a teardrop hired for specific jobs requiring a particular capability. Cass had heard that the teardrops could only die if they shed their last tear. She wondered if the story was true and if she should ask Tor about it. While she was wondering about the myth, her partner observed the people around them. His eyes kept glancing back and forth from the window to the other customers while occasionally sipping the beer. Tor looked stoic as ever, the contrast of bored, sad, angry, and alert all in one. Underneath his cloak was ragged clothes, and the leather harness across his chest held several weapons.

A minute later, the table shifted slightly, snapping Cass out of her reverie. She glanced in front just in time to see a man sitting across from her. Cass shook her head, grabbed her black braid, and leaned back in her chair. Tor took a long drink from his beer and looked sideways at the man. They both observed the new guest at their table. The stranger looked to be in his late twenties. He had long blonde hair, blue eyes, and tattoos that covered his pale arms.

The man grabbed both sides of the table, looked at Cass, and raised his voice, "I need more time! I will get the money."

The partners were silent; they didn't utter a word as if he hadn't said anything. It was all usual for Cass and Tor; whenever it was time for recovery, the contractor would beg for more time. She knew it was all a facade to show that he wasn't afraid of them. Nonetheless, his trembling legs and the fear in his eyes gave him away. Cass smirked. *Not so intimidating, huh?*

"Do you hear me?" he shouted, leaning forward.

Tor quickly sat down his mug of beer on the table, frustrated. The man glanced at Tor and, intimidated, immediately sat back in his seat He realized that Tor was not someone he could mess with. Rolling her eyes at the man's tantrum, Cass leaned forward and took matters into her hands.

"Sir?" she started saying. "I can't give you any more time; that is not what we do." She gestured to Tor. "We collect debts, and yours is due."

Startled, the man looked at her and then at Tor. Anyone could see how terrified he was at that moment. The partners waited calmly as if already knowing what his next move would be. And they were right; before anyone could move, the man pushed the table toward Cass, leaped up, and turned to run. However, he wasn't as fast as he thought. The man had only taken two steps when Tor grabbed him and pulled him back down to his seat. The DNA hunter above hissed and moved closer. Cass glanced sideways at the DNA hunter and shook her head. *The debt needed to be collected today or else...* she thought, sighing heavily.

Cass leaned across the table, "You owe Mom an arm, and there is no way we can give you more time now."

She pulled a small bag out of her pack with a compression freeze bandage and a frozen bag. She looked up at Tor; he nodded and, in the blink of an eye, separated the man's arm from his shoulder with a short sword. The man cried in pain, the crimson blood oozing from the wound. Before he ran out of blood, Cass applied the bandage on his shoulder. Her partner, Tor, cleaned up the arm and placed it in the frozen bag.

"Well, looks like your debt is paid," she said, smiling. "Thank you for cooperating with us."

The man scowled in pain and tilted his head down.

Tor tapped him on the shoulder and said, "You can go now." He shifted the bag on his shoulder and nodded at Cass. Together, the partners walked out of the café, leaving the DNA hunter hissing behind them.

***

The brightness glinted off as an oval-shaped machine, the fog of steam enveloping it. The machine's upper body looked like a sea manta covered in filters and antennas, which used people's genetic code to track and hunt them. The device 'DNA hunter' hovering over a small café left and flew quickly between the buildings and overhangs.

After an hour, the DNA hunter buzzed and stopped at the dark alley. Three people in the hood and a man that radiated authority were already waiting for the hunter. When the man learned the hunter had failed his mission, he shot a blue beam at the machine. The hunter whizzed and fell, splitting up into small pieces.

"Useless!" the man shouted, making the other three people cower back in fear. "They were able to collect the debt because of your negligence!"

"I—"

One more beam of light shot through the dark alley, and a hooded man fell on the floor. The remaining two looked at each other and backed away from the man.

"Do you want to say something too?" the man growled. The hooded figures shook their head, and the man nodded. "Good. Pick both of them and let's go; we have to move forward!"

Chapter 2:

The Broken Hearts Service Company

The automatic door slid open, and Cass and Tor walked inside the company. Walking to the left side of the building, they swiped their ID card on a small machine. The red blink was all they needed to know to proceed further. The front gates of the company opened up to a large area where small counters were being operated for different work.

The Broken Heart Service was buzzing; familiar and unfamiliar people moved around the corridors in a frenzy. Half the people moved through the crowd with body parts in freeze-dried bags and clipboards. While the remaining were standing in line at a counter.

With bags in his hands, Tor made his way to the recovery department, with Cass following closely behind him. But she fell back on her step when she heard a loud noise. Looking toward the right side of the building, Cass saw a man in his late thirties arguing with the girl at the counter. A few security guards surrounded them, trying to calm down the man.

Confused, Cass looked at the counter, 'new prosthetics,' and sighed. *Here we go again!* She thought. The service company didn't only loan money, but they also sold or loaned body parts and prosthetics, but the transactions were done elsewhere as this building only dealt with contracts.

"Cass?"

Cass looked in front to see Tor calling her.

"Yes, I am coming!"

***

Fifteen minutes later, Cass and Tor entered the recovery department on the fifth floor. The department was large, with ten cubicles, three offices, and a small separate café. As Cass walked through the cubicles to her office, she could smell a mixture of two distinct smells, medicine and cologne. The main hall of the department was lit by the glow of four small, unshaded lights hanging from the ceiling in the center of the room, and the three floor-length windows added more brightness to the room. The design of the office was simple but chic, white and dark grey.

Even though it had its fair share of enemies, Cass thought the job was good and had its adventures.

Once in her office, Cass began to file the paperwork. The partners had to fill out a procedure to place the body part they recovered in the system. Cass didn't need to look back to know that Tor followed her into the room, moving around, taking in everything.

As if he didn't spend his whole day here, Cass thought, shaking his head.

"Shit!"

Tor's loud voice, followed by an exasperated sigh, startled Cass.

"What—" she halted as she caught sight of an elderly lady, Mom, through the front office and quickly realized the reason for Tor's uneasiness. She chuckled slightly when she remembered how much Mom loved teasing Tor.

Everyone at the company called her Mom or Mistress of Medicine. She towered over half the people in the office, including Cass; Mom was haughty and had sharp and sultry features. Her walk was graceful, as if she was floating on the ground. A powerful independent woman who was intelligent enough to know when to be logical and wise. Cass had heard rumors that resolved around Mom; it is usually said that 'she could steal anyone's heart if she wanted to, but everyone gave it to her on their own.' She was a stern woman that ran a tight ship. When she entered, every worker seemed to pick up their pace.

Tor grimaced when Mom made eye contact with him. Cass tried to hide her smirk when Mom reached them. The relationship between Mom and Tor was a complicated one; the more he struggled, the more mom enjoyed tormenting him.

It was a tug-of-war, and Cass was a peace barrier between them.

"Tor!" Mom exclaimed and latched on to his arm. She leaned forward, pretending as if she was going to kiss him, and as usual, Tor leaned back. Cass knew that Mom's gesture made him uncomfortable, but not so much to push her away. Mom never took his reaction to heart and always smiled at him.

Once she was done playing with Tor, Mom turned toward Cass and said, "Good job! I knew I could count on you two."

"Thank you," Cass replied while Tor rolled his eyes.

"Alright, I have another recovery for you both," Mom told them, never letting go of Tor's arm. She pulled a red file from her bag and placed it on Cass's desk. Tor reached for the file only to have Cass grab it quickly. She went through the assignment and passed it to Tor.

"We get to take the train to the countryside," Cass informed him. She always got excited whenever they got to do an out-of-town job, as she loved traveling by train.

"Alright!" Tor nodded and started to move forward, but Mom's stronghold ceased him from doing so. Tor looked at Cass with a lost expression, and she smiled at the teardrop. *A man who can face several armed men and has seen his share of combat in the old wars is helpless in front of a woman.* Cass thought, *just like a mouse caught in a cat's trap.*

"We will leave now," Cass told Mom and grabbed Tor's free arm. "Come on, Tor! We can't be late for the train."

***

At a far place from the Heart Service company, a man paced back and forth in a dark room. There was no light or sound except the occasional thunder that lit the room. A group of people watched cautiously and waited for his order.

"There is a new assignment," the man in the shadow snarled. "I hope you all won't disappoint me this time."

"We won't, Sir," the group exclaimed.

"This time…this time, I won't let them go!"

Chapter 3:
The Train

Three hours later, Tor and Cass arrived at the train station and at the platform. The rail system was the fastest way to travel as the modern train used electromagnetic energy. It was easier and more feasible for the duo. On the way to the station, Tor had already booked their tickets. And once there, Cass went to the ticket counter to get their tickets. She stood in the line and waited for her turn, scanning her surroundings.

It's been so long since I have traveled on a train, Cass thought, smiling. She always loved traveling on the train and felt pleasant when it floated on the rails—quiet and rapid.

A few minutes later, she took her tickets and went toward Tor. Without a word, Cass stood beside Tor and waited for the train. Sighing, she looked at Tor and noticed he was looking on the other side of the platform. Curious, Cass followed his gaze to the other side only to roll her eyes.

On the opposite side of them were her worst rival—Mods.

Six mods roamed near the benches, nervous and anxious. They sold their arms and legs for modified body parts and usually worked for corporations as mercenaries or hunters. Every time Cass saw them, she felt a pang in her chest. Cass had a modified part in her body; half of her heart was mechanical.

Cass was clueless as to why her heart was in that condition; she couldn't remember anything about her past. Her heart was powerful and robust, and she used it to power her weapons. Every weapon that was in this world was powered by the pulse of the person using it. So, people that had mods have the modifications to make their weapons better. So, all weapons were percussion weapons or bladed weapons.

Tor has no mods, Cass thought. *But I guess he doesn't need them.* Teardrops were trained traditionally; they didn't need percussion weapons to fight; they were skilled. Cass didn't know much about her partner and his past. She wanted to learn more about him and sometimes wondered asking Tor how he became a teardrop.

"Looks like we are going to have some competition for this recovery." Tor's amusing tone snapped Cass out of her reverie. Peeking at him, she saw him staring at the mods with a little smirk. Cass did not care about them; to her, they were a nuisance.

A second later, the train pulled up, and Cass shouldered her small black bag.

“Come on, Tor!” she exclaimed, looking at him sideways. “It’s time to go.”

***

For an hour, Cass and Tor rode silently on the train, minding their business but were alert and attentive to the surrounding. A click sound forced Cass to look up from the window to the mod. She watched as he walked past her and Tor to the other car. Perplexed, Cass looked farther up the car and noticed the mods sitting three seats down them. Every so often, one of them would get up and walk to the car behind them.

Doubtful, Cass nudged Tor and asked, “What do you think they are doing? They keep going to the other car.”

Tor glanced up from the book he was reading and turned around. He observed the mods and said, “They probably have a DNA hunter in the car.”

When another mod moved to the car behind them, Cass got up and followed him. The moment she entered the car, she realized that Tor was right as a DNA hunter floated in front of her. Cass was always fascinated with them and curious when one was close by. With several antennas on the back, and the front shaped like a manta ray, the hunter looked enthralling to her. Cass watched the filters that would come up whenever it flew around.

The filters strain particles, detect DNA to track people down, and transmit the location to the mods, she thought, watching it closely.

The DNA hunter arms were wide open and shaped like an oval, hissing steam out of its vents.

Lost in her thought, Cass didn't notice the door opening. It wasn't until she heard the click sound that Cass looked up and saw the mod.

"Step away from it!" the mod grunted, glaring at her.

"No problem!" Cass said. "I was leaving anyway."

Standing in the doorway, the mod leered, "I don't think so! No way you two are getting the recovery on this one."

The mod raised his arm, a percussion weapon appearing in place of his hand, and struck forward. Cass ducked down, and the wave went over her head. She quickly drew her weapon, and the pistol fitted perfectly in her hand with the magnetic tubes connected to her bracelet. Within a second, she charged the weapon. The mod sent one more blast in her way, and Cass dodged it too.

And in a swift moment, she leaned up and hit the mod in the chin. The mod stumbled back from the impact. Cass took the opportunity and leaped forward, putting her weapon to the mod's chest and firing. The powerful percussion knocked the mod back against the wall, and he fell down, unconscious. For the last time, Cass let her weapon charge again and shot the mod's arm. She knew that it would disable the mod for the time being.

I need to get out of here, she thought, peeking around her.

***

Cass flung open the train door and crouched down with her weapon ready. She was prepared to lunge forward and take down the mods. But to her astonishment, she didn't see any danger and straightened up. Moving forward, Cass noticed Tor and the mods sitting in the same spot. She walked up thinking all was well and nobody noticed anything. Slowly she sat down and was about to question Tor, but before she could, the mods across from her got unconscious, and several mod parts appeared in their seat.

"Did you find the DNA hunter?" Tor asked, turning away from the window.

Cass nodded. Tor didn't say anything and looked back out the window.

Chapter 4:

The Festival

Three hours later, the train arrived at the station, and Cass and Tor exited their car. Tor threw the bag filled with mod parts on the train floor as they stepped out of the train. The mods limped and walked over to the discarded bag, and picked up the parts. Cass glanced back at them and shook her head.

The parts were like a drug to the mods, she thought and shook her head.

After getting attached, a prosthetic helped the body produce a chemical in the brain. The mods were addicted to the chemical, so the more mods, the more chemical would be produced. Cas turned away and quickened her steps to catch up to Tor. He didn't care about them and walked away, ignoring them. The duo departed for the village where their next assignment was.

The duo soon entered the small mountain village, where they observed the town lightened up in a festival.

Glancing around her, Cass saw the banners strung on the buildings and small pop-up shops set up one after the other that sold local goods. She could hear soft music coming from what she presumed was the town center. As they walked around the village, people regarded her with nothing but a smile and a nod, but they treated Tor differently. It felt like everyone almost bowed to him. Granted, Tor was at least a head taller and bigger than most people and walked with a sense of purpose. People—children or adults—nodded and just moved out of his way.

"Tor, why is everyone so weird around you?" Cass inquired.

Cass watched as Tor's expression turned annoyed, and he grumbled, "The Teardrop camp is nearby, and the people here have been the main support of the camp for generations. Whether it be good or bad, a teardrop in town during a festival is…a sight." Tor looked away from her and started walking a little faster." This way, the address for the recovery is up ahead."

"Hey Tor, we should stay and enjoy the party?" Cass commented, looking at the festival around her.

Tor huffed and walked away, pretending he didn't hear her. Cass rolled his eyes, knowing that he was being grumpy, and followed him.

***

Half an hour later, Cass and Tor walked up to the house. Immediately, Cass stepped on the porch and knocked on the door. Tor stood back as always, waiting for something to happen. Cass was prepared for the usual—begging and offering anything—to get more time to pay the debt.

Most people would take the loan, and when they weren't able to pay them back, they would hide in the mountain villages to get more time. The door slowly opened a middle-aged woman appeared in front of them. The woman looked tired and wrangled, her eyes red from crying. Behind her stood a boy who looked to be around ten or twelve years old.

Before Cass could speak, the woman intervened and said, "I know why you are here…but you are too late." She paused and looked down." My husband passed away last week."

Here we go again! Cass thought, sighing. She had heard this before and knew the man was hiding in the house or somewhere nearby. She rubbed her face and glanced at the young boy behind her. He stared past her at Tor, who stood just off the porch.

"Is that a teardrop?" he asked, tugging his mother's arm.

Startled, the mother pushed him behind her, hiding him from them. Cass looked at them closely, trying to understand their behavior. Sighing loudly, Tor stepped forward and pushed past Cass to the women. He forced open the door making the women step away. Tor walked inside the house and searched for the man but was unable to find him. Frustrated, he stepped out of the back door with Cass and the man's family in tow.

In the backyard, on an elevated wooden platform shrouded in drapes, flowers, and small trinkets, lay a man in his mid-50s. His body seemed distorted from pain and suffering from a long illness. Cass realized what was going on. The man knew he was terminally ill and took a loan in exchange for his legs. So now his family would have money to survive.

She looked at Tor and said, "The debt still needed to be paid. The mods and the DNA hunter will be here soon. Someone has to pay." Cass glanced at the small boy standing in the backyard. Tor followed her stare and immediately spun around. He kneeled down and looked at the boy.

"I will pay my father's debt," the boy said firmly. "So, my mama can be okay."

Tor grabbed the boy by the arm and picked him up; the boy remained defiant, staring back at Tor.

“We must move fast!” Tor told Cass, turning around and looking at her.

Tor spun the boy on his back and sprinted out of the backyard. He ran up to the hillside deeper in the mountains. The mother cried and fell down on her knees; Cass looked at her and sighed. She didn’t have time to say anything, so she ran ahead, catching up with Tor.

Chapter 5:
The Mountain Hunt

In the sunlight, the mountain peaks contrasted different shades of gray, ranging from sweet blue-slate to silver-white. Cass followed her partner, Tor, and ran up the mountain. Nobody said anything. There was complete silence, and only the sound of their boots clapping against the rocky pathway could be heard.

Tor had the young boy on his back, and a leather strap around his waist tied the boy to the harness he wore. The teenager's feet were placed in small stirrups, which allowed Tor to move freely without holding onto him.

"Tor!" Cass yelled. "What are we looking for or doing? I cannot understand what is on your mind!"

"Something," Tor muttered. He didn't stop and kept moving." We need to get higher up the mountain."

"Why?"

"You will know when we get to the top of the mountain."

Cass sighed. Even though she couldn't comprehend what was happening, she believed in him, so she followed Tor quietly.

***

After several minutes of climbing and running, Tor and Cass came to a small outcropping of rock, looking across the valley below. Confused, Cass glanced around at her surroundings; she could see the village below starting to light up for the festival.

We need to hurry, she thought and turned back to Tor, who was kneeling in front of the boy.

"What is your name?" he asked the teenager.

"Ben," the meek boy replied.

"Alright, Ben, this is going to hurt," Tor told him." I want you to be strong right now, alright?"

Ben nodded and peered at Cass. Cass sent a small smile his way to help him calm down.

Swiftly, Tor pulled out his knife from under his cloak. Cass gasped, startled. She knew that Ben would have to pay his father's debt, but the sudden moment shocked her.

Is Tor going to take his legs right here? she thought and winced slightly. She was about to say something when Tor looked at her.

"You need to configure your weapon to rifle mode," he told her.

Still perplexed, Cass pulled her weapon out. Leaning down on the hard rocks, she took the parts out of her pack and started to make the rifle. As Cass was putting it together, she glanced at Ben and Tor. She noticed that Tor had cut Ben's hand and was capturing the blood in a small metal vial. Once the vile was full, he handed Ben a rag to clean the wound.

Tor walked over to Cass and kneeled beside her. She had the rifle together and was waiting for his order. Tor motioned toward her bracelet. Cass immediately understood and aligned her left hand with his. The partners had identical bracelets on their wrists. It was black and silver, programmed to charge the weapon

Tor pulled one of the tubes out of the bracelet on his right hand, and they magnetically connected with hers. Cass could see the heartbeat monitor charging on her bracelet. The bracelets, when connected, combined Cass and Tor's heartbeats into one and, in turn, helped charge the weapon.

"Here, take this," Tor said and handed her the small projectile with the blood in it. He pointed to the ledge some hundred meters away. "Do you see the mountain goat?"

Cass followed his gaze and saw the all-white, hoofed, big mammal-eating grass. She nodded and said, "Yes!" Her eyes widened when she realized what was going on.

Oh! So that's his plan! She thought.

She put the vile in the barrel and straightened up, shouldering the rifle. They listened as the beeping grew stronger and powered the weapon. Cass took a deep breath and fired, aligning the weapon in the direction of the goat in her sights.

The magnetic field hit the goat, and it stumbled a little, causing it to stop eating. Suddenly, he backed away and sprang away up the mountain.

"It will buy us some time," Cass said, and Tor nodded.

He looked back at Ben and said, "Okay, listen to me, run back to your home, take your mom, and leave the village tonight. Go back to the valley. Okay?"

"Okay!" Ben said and nodded. He folded the rag around the cut and climbed down the mountain.

Tor looked back down the ravine to the village and muttered, "Well…now that we are here, we might as well enjoy the festival."

"I like the idea," Cass said, smiling.

Cass and Tor slowly walked back down the path they had earlier mounted. A few seconds later, Cass heard the familiar sound of a DNA hunter. Curious, she glanced back to see it flying through the ravine and scanning the hillside. She watched the hunter climb higher into the mountains and turned around, smirking.

Chapter 6:
The Campfire

The festival was in full splendor when Cass and Tor entered the town. The townspeople were singing and dancing. The people had set up shops and were reaching out to the visitors. Some of them marketed their products by boasting about them; other people used flags to get people to purchase their goods.

Cass walked past the shops, looking at trinkets and all sorts of oddities. She glanced back at her shoulder and didn't find Tor there. She blinked and then shrugged, going back to what she was doing. Cass wasn't worried about where he was; she was sure he was looking for a tankard of ale.

"Something like that," she mimicked his gruff voice.

Cass continued to walk around the fair, taking in the sights. She was engrossed in the activities going on around her, smiling and laughing at the entertainment. She was lost in her thoughts and didn't notice the intruder until she heard the distinctive click.

She was alerted when she realized the moderator was walking behind her. Cass slightly turned around and peered at him from the corner of her eyes. The mod passed by her, not noticing her. He was heading out of town and up the slope.

I wonder how long they will chase that mountain goat. Cass thought, grinning. *Hopefully long enough to get the family out of town.*

"Hey!" Tor said, walking toward her." How about we go and sit by the fire?"

"Sounds fun," Cass replied.

"Let's listen to stories and eat something; I am starving."

Tor led the way, and Cass followed along. At the festival, there were many storytellers gathered around the campfires, telling old tales. Their audience sat on the log, ate, and listened keenly to the tales. Cass found an empty log beside Tor and sat down. She sighed and glanced around at the people. In the corner, she noticed a couple of kids eyeing them. Cass smiled and motioned them over to sit and eat with them.

After a moment of hesitation, the children sat beside them. Tor grunted but then smiled, passing on his cake to them. Cass, Tor, and the children were in the middle of a conversation when a storyteller jumped in and started to tell a story.

The man was loud and had a habit of using too many gestures to recount his story. He told some funny stories about a goat, a pig, and a fat innkeeper. People laughed and cheered him on. Then the storyteller went to his next story—the story of the first teardrop.

Cass put her paper plate beside her on the ground and leaned forward. She wanted to hear the story. Meanwhile, Tor paid no mind; he was in a serious thumb wrestling contest with two of the kids.

***

"The tale of the mighty first teardrop started at the dawn of time and before mighty cities were raised from the earth." The storyteller started moving his hands around. "This warrior lived peacefully in these very mountains after he had been in a great war that carved the entrance into the valley. The war emanated from the valley." The storyteller paused for a moment and moved closer to the fire to make it more theatrical.

"The invading army marched across the valley, trampling all in its wake." The storyteller began stomping, trying to illustrate the scene to his audience. Cass listened intently. "Before the massive army could get to this very village, the warrior came down from the mountain, with sword and spear in hand, and riding atop an enormous mountain ram. The ram snorted as it walked, and off its curved horns were bells that jingled as it moved. The warrior met the army in the field and taunted them, making them chase after him. The warrior never got too far away and guided them away from the village. Suddenly the warrior heard a cry, and he stopped. When he turned around, he saw his lover atop a horse riding alongside the army. He turned toward them, fearing for his lover, as she was in danger. The warrior bravely fought with a spear, a sword, and his hands. Unexpectedly, he fell from his mighty ram, and the bells didn't jingle anymore. In his haze, he saw his lover walking toward him. She kneeled in front of him and grabbed his sword, thrusting it into his heart."

Silence fell on the crowd. Everyone was engrossed in the tale.

"The mighty warrior fell as his last tear rolled from his eye. With his heart broken, the tear fell to the Earth." The storyteller paused again and added drama to the story by throwing another log into the fire. "Up from that tear!" he yelled, triumphant. "A well sprang to the Earth. The water was so powerful that the army ran away in fear." The storyteller glanced at everyone. "The mighty warrior died, but on his death, the well rose up. Some say that if you listen closely, you will be able to hear the bells. It has been said that the mighty warrior patrols these hills."

The storyteller kept ringing a small bell while he stepped away from the fire.

Cass leaned back and looked at Tor. He was sitting calmly with the kids. Catching his eye, Cass smiled at him, and in return, Tor grinned at her.

"I was told he rode a dragon," Tor said cheekily.

Chapter 7:

Mr. Blight

Cass was sitting in the dining car of the train, enjoying her lunch and thinking about the festival. After staying for an hour, Cass and Tor decided to leave. They had to go back and let Mom know about the payment. She kept wondering what they would tell her.

Suddenly, she heard the telltale-clicking sound of a mod. Alarmed, Cass looked up from her breakfast to see a moderator walking up the aisle. The clicking sound was always apparent. Cass turned a little in her chair and shifted her weight forward. She stayed still and waited. To her surprise, the moderator walked past her and sat at a table across from her. Cass regarded him with a sidelong glance; it was the same one she had exchanged in the railcar a few days earlier.

I need to be alert, she thought.

Cass shifted a little on the side so that it was easier for her to grab the weapon tucked in her shoulder harness. She pretended to be aloof and purposely picked up a slice of apple. She popped it in her mouth and chewed softly. A minute passed, and the mod didn't move or show any sign of attacking. When it was clear that he meant no harm, Cass disregarded him and returned to her meal.

When she was about to take her next bite, Cass saw the moderator grin at her. Irritated, Cass sprang up from her seat and moved toward him; however, a tall man interrupted her.

Without any warning, the man sat at her table across from her. Cass looked at him closely, wary of him. The man was dressed in a black dress shirt and pants that were paired with a long brown coat. His dark brown eyes were mesmerizing, but they had a distant and cold look to them. His short dark curly brown hair was well-groomed and well-kept.

"Hello, Cass!" he greeted and motioned her to sit down. "I am Mr. Blight, from the Blight Contract Company."

"Hi, Blight!" Cass replied with a gleeful smile. "How are you?"

She sat down and leaned forward in her seat. Cass knew that he was probably the boss of the mods and the DNA hunter that the partners sent searching for the mountain goat.

"I am well," Mr. Blight said, chuckling." I like the competition and the little goose chase you and your partner sent my men to."

“It happens.” Cass smiled.

Blight shrugged and said, “I am more interested in the weapon you have.” He motioned toward her bracelet, and Cass shifted slightly to grab her pistol.

“Thanks, I am glad to have it,” she said.

Mr. Blight stood up and placed his hand flat on the table.

“Make sure you keep your heart safe,” he said, smiling and staring at Cass darkly.

Cass didn’t back down and stared back. Softly, she tilted her head to the right making her braid fall over her shoulder.

“I certainly will, Mr. Blight.”

Mr. Blight nodded and pressed his hands on the table. Abruptly, a percussion wave lit up from the ground and shattered the table. Quickly, Cass jumped up, pulling out her weapon, but by then, Mr. Blight was walking away with the mod behind him.

“Not today, little one!” Blight uttered, waving her off and exiting the dining car.

Chapter 8:

Mistress of Medicine (Mom)

Cass stepped off the train and noticed a sudden frenzy of movement around her. Boxes were being stacked, and people were milling about in aimless directions.

What's going on? She thought to herself, all the while searching for someone familiar. Finally, she saw the reason through the line of chaos.

Mom.

Cass's mother stood in the middle, navigating the crowd. She was giving orders and overseeing the shipment of parts being transported to another city. Cass noticed she was wearing pants, which was odd, provided that she would usually wear a nice dress for walking or running errands.

Hmm, something's amiss. Where is Tor? She turned around, moving her head to and fro in order to locate him.

No sooner than she finds herself submerging in the ocean of panic, the sight of Tor immediately lent her a sense of calm. He was standing with other teardrops talking. The large red lipstick stain on his cheek told her that Tor had already had a run-in with Mom. Cass laughed and walked up to Mom, who was coaching six people on what to do. She stood back and watched them run off; happy they were given a task.

What even. Cass rolled her eyes and shook her head.

She knew it was absurd, but then again, who wouldn't react this way? Mom was a striking woman for her age. Long black hair pulled back in a braid, gyrating amid her poised walk and soft doe eyes exuding a sharp intensity to them.

Mom caught sight of Cass and turned to her.

"So, the contract had passed away, I assume," Mom questioned.

Cass nodded. *She already knows.*

Mom pointed to Tor and said, "He mentioned something about the debt all men are supposed to pay. So, I assume no payments were made." Mom sharply looked at Cass. "I am in this business to make money and fill contracts. Your little adventures are costing me a lot. I could fire you right now and get rid of Tor if only I didn't have so much invested in you."

Cass stood there, knowing that Mom knew what had happened. She waited for the wrath in tow but was surprised by her reply.

“But I guess that is what happens when a teardrop is involved,” Mom said and smiled. Her face softened a bit.

Cass nodded.

“Very well, we will move on, you and,” Mom motioned to Tor. “Owe me a contract. There is a meeting today with a man who wants an extension. I want a token from him.” Mom turned around and moved away, giving orders to others.

Cass sighed in relief. She glanced at Tor to see that he was already grinning back at her.

Chapter 9:

The Sniffing Man

A few hours later, Cass sat in the café, which happened to be her favorite among other eateries. The place had mouth-watering food, a well-cooked burger, and an amazing view out of the window. Her partner, Tor, sat at the table beside her, quietly nibbling on his food. They both enjoyed their meal while waiting for the contractor to show up at the meeting.

A man walked up to the table and sat down across from them. He sniffed loudly and said, "I am here. Are you the workers from the broken heart company?" He sniffed aloud again.

Cass tilted her head, looking at him, and was about to say something when the man sniffed up again and said, "I need an extension." He paused. "I need my arm. I will give you the money in two weeks."

Cass beheld him again and was about to speak when the man sniffed again. This time she raised her hand, stopping him. She sighed and told him, "You know we will have to stall a DNA hunter to make time for you?"

The man sniffed again, and she continued, "Mom said she wanted a token. Do you have one?" Cass looked at him sternly.

The man peered at her and sniffed again. Before he could speak, Tor grabbed his hand and pulled it onto the table. Swiftly, Tor sliced off his ring and pinky finger. Cass blinked, surprised, and then glared at Tor.

Tor glanced back and said, "You were taking too long, and his sniffing annoyed me." He turned to the man whose eyes were wide and mouth hung open in surprise. "This token will do." Tor pointed to the fingers. "Go home, close your window and house locks. We will take care of this."

Cass gazed at the man and said, "If you do not pay us back in two weeks, we will come back for the rest of the arm."

Tor grabbed the fingers and placed them in a pouch. He looked at Cass and nodded to her. He walked off, leaving them alone. Cass stood up from her seat and began following her partner. Patting the sniffing man on the shoulder, she said, "After two weeks, we will be back."

Cass walked out of the café and next to Tor.

"A little grumpy today?" she asked.

"That sniffing ruined my lunch," Tor grumbled at her.

Chapter 10:
The Park

An hour later, Cass was lying in a prone position atop a small hill in the park. Tor and Cass would have to take out a DNA hunter to slow down the contract. Hunters were programmed with the contractor's DNA. When the contractors took the loan, they all provided a DNA sample, so when they couldn't pay back, the DNA would be loaded into a hunter, making it search for them. A hunter needed to be immobilized for the extension to work. Depending on the contract, it would take about another week or so before another DNA hunter was programmed.

Tor and Cass were baiting a hunter with the sniffing man's fingers. They chose the park to limit the hunter's movement, with tall trees around them and paths where they could ambush the hunter easily. Cass had her weapon in rifle mode and already attached to her bracelet. With every heartbeat, the weapon started charging, and the velocity increased. She looked down the path they had chosen. She could barely see Tor kneeling beside a big tree with his cloak over him.

Cass scanned the sky for the hunter. *It's a waiting game now,* she thought.

Instantly, she heard the swoosh and hiss of a slow-moving hunter above her head. Cass hoped that there was not enough DNA of the sniffing man on her so that it might track her too. She rubbed her hand slowly in the grass, hoping it would mask the DNA. The hunter continued past her, dropping into the pathway. Tor had hung the bag with the sniffing man's tokens on the tree branch.

The hunter hovered over it, circling it as if teasing its hunt. As soon as the filters in front of the DNA hunter were in view, Cass fired a percussion shot. The hunter hissed like a snake and slanted downward. Cass shot again, but this time there wasn't any power in the blast, making the hunter move a little.

Tor leaped up from his hiding spot and, throwing his cloak over the hunter, pulled it closer to the ground. Jumping on top of it, Tor tried to weigh it down. Quickly, he used his piked ax and cut off the hunter's antennas. Tor raised his hatchet to strike again when he felt a shot go past him.

Surprised, Tor looked up at Cass, and she pointed behind him. Cass caught sight of a mod sneaking behind Tor with a weapon and shot him before he could strike at Tor. Alerted, Tor spun around to see four mods running up the path toward him. Tor rolled off the hunter and landed on his feet, leaving the cloak on the semi-disabled hunter. As soon as his feet hit the ground, he pulled out his short sword from under his harness. With a piked ax in one hand and sword in the other, Tor lunged forward. He attacked a mod in the chest with his body while thrusting his sword at the mod's mechanical arm. He moved the sword outward, causing the arm to separate.

Tor heard another percussion wave and looked back to see Cass shooting the hunter again to keep it from flying off. He spun around and moved back as a blast from Cass's weapon hit the next mod in the chest. Tor grabbed the mod's head before he could fall back and threw it into the ground. He used his ax and cut the mod's left leg off. Taking another step, he pivoted on his front foot and stepped upward, smashing his fists into the next mod knocking him down. Before he could react, Tor moved again and sliced the mod's arm.

Another blast hit went past Tor. He didn't turn around now, knowing Cass had reshot the mod or the hunter. He glanced around and noticed one mod was missing.

"Tor!"

Tor heard Cass yell and gazed up at the hill where she was.

"The mod has the hunter, and he is running away!" she shouted, pointing her finger in their direction.

Tor took off in the direction Cass pointed. He saw the mod dragging the half-disabled floating hunter in front of him. Tor picked up the pace and, within a second, covered the distance quickly. Cass shot again, and the mod jumped forward from the hit but kept running.

The mod has modified legs, so it moves relatively fast. Tor thought, grimacing.

Running, he grabbed his ax and aimed it at the mod. Unluckily, the ax missed and struck a tree. When he ran past the tree, Tor seized the ax. Another blast went past him striking the mod again. This time the mod dropped the hunter. And with its modified legs, he took off even faster.

Tor stopped beside the hunter that was still barely hoovering and watched as the mod ran off. He attacked the hunter and pushed it to the ground. Lifting his ax, he knocked it out.

Cass caught up to Tor and said, "It isn't good, is it?" She gazed down the path and around at the mods scattered on the floor. She handed Tor the fingers, noticing that he was glum. "What's wrong?"

“We need to get Mom her token quickly,” Tor said and shook his head. “We need to make some plan to get out of sight.”

Cass nodded. “We have a lot of run-ins with mods lately.”

“At least Mom gets her token,” Tor grumbled.

Chapter 11:
Out of Town

An hour later, Tor handed Mom the token. Cass watched as Mom tormented him, despite his willingness. She continued to touch him and smile seductively at him. Cass shook her head at them and smiled lightly. She knew it was a harmless quarrel between them and nothing else.

Tor looked at Cass and said, "We are going to leave for a while. We are catching too much flack."

Cass shrugged and wondered what 'flack' was. Mom just smiled and started to glide away.

"The countryside is always a good option to hide for a few days," Mom commented.

Tor grimaced as she walked away and turned to Cass.

"Go pack up some things; we are going to the mountains for some time," Tor instructed her.

"Alright," Cass said and nodded.

"Meet you at the train station in an hour."

Cass watched as Tor walked after Mom and stopped her. They started having a conversation, which got pretty heated within a minute, leaving Mom seething. She huffed but nodded and stomped her way out of the office. Cass sighed and walked over to Tor. Patting him on the shoulder, she smiled at him and glided away. Cass knew then that Tor must have told Mom they would be gone for more than a few days.

***

Half an hour later, Cass walked out of her apartment, grabbing just a few things. She took only what was necessary for the journey and avoided extra baggage. *He will be waiting for me at the platform,* she thought. When she arrived at the train station, Cass saw Tor standing on the platform like he had been waiting for her for hours. *Knew it!*

Cass walked up to Tor and said, "Are you ready to go?" She didn't wait to hear his answer and walked past him getting on the train.

"Umph."

Cass heard behind her, knowing it was Tor, telling her 'yes' in his unusual way.

Once seated, Cass glanced at her surroundings. When she was satisfied that there wasn't any threat around her, she looked at Tor. He was sitting across from her, staring into nothingness.

"Where are we going?" Cass asked.

"To the village we last visited," Tor answered.

Cass knew he was going to check on the boy.

"Didn't you grow up in the mountains?" she asked. "Why can't we go there?"

Tor looked at her and said, "I haven't lost anything there, so we don't need to go back."

Cass nodded and settled back into her seat. She shifted until satisfied with her position and let the train's rumble lull her to sleep.

***

Cass squinted her eyes as she stepped off the train. *Where are we?* She thought and looked around, confused. *Did we stop at the wrong location?*

"Tor," she called and peered at him. "This is not the village platform, is it?"

Tor shrugged her off and said, "We are going to stop here for a while before we move forward to our destination."

He pointed to a gravel road that winded into the mountains. Cass caught sight of a wooden sign on a tree with 'camp' written on it and an arrow pointing in the direction of the path. She jumped as a loud whistle echoed in the silence, and the train moved away, leaving her and Tor on the small platform.

"We are going to the teardrop training camp," Tor said and turned around. "You need to stay out of stuff. Don't touch anything, be humble and polite."

Cass looked at him sideways and grunted, trying to imitate him. They got hold of their small bags and began walking up the path.

"Tor," she asked. "Why did you become a teardrop?"

She waited for a lecture or a stoic look. But to her astonishment, he did answer her this time.

"To live as a broken heart and to have a purpose," he said plainly, looking far away into the distance.

Cass could feel something more, but she said nothing and just walked silently beside him.

Fifteen minutes later, Cass walked through the teardrop's camp open courtyard with Tor. She saw men training in combat skills, others reading in outdoor classes, and some cleaning. Every so often, one of them would walk toward them, nod at Tor, and quickly glare at her. Some would say something to him and move on.

Cass didn't know why they behaved this way, but she knew better than ask her partner and quietly followed Tor into the main hall.

"Tor are there any female teardrops?" she asked.

"No," he answered. "You must remain here in the main hall. Do not touch anything, and be patient. There is a small teardrop ceremony for men who have finished teardrop training. So, wait here."

Tor turned and walked off, leaving Cass in the hall of honor. She watched him walk off and glanced around at the hall. She looked at the statues, paintings, and murals situated or hung all over the place. She observed the pictures that depicted what she guessed were great teardrop battles.

She was fascinated with the stories behind these paintings. One was a picture of a man with a ram's head in his hand with bells on the ram's horns. Cass chuckled, remembering the storyteller's story. She looked at the painting of the first teardrop who died in battle and shed his last tear, making a well spring up.

Such silly stories, she thought and chuckled again.

Thud. Thud. Th—

Suddenly, she heard the sound of beating drums coming from the end of the hall. Cass walked toward it and peeked through the crack in the door. She saw a great granite fountain with steps leading into the water.

Around the fountain stood hooded men playing drums. They were leading three men dressed in robes to the fountain edge. The men in robes had their heads down and walked obediently as the drums played in the background. At the edge, they all stopped, and the men in robes took their hoods down. The sound of the drum got louder and more prominent by then.

An older man stepped out of the group, still seeming healthy, and moved with prowess. He walked waist-deep into the well. Cass noticed Tor standing along the fountain edge. His face had a certain calmness as if it was a normal routine. Swiftly, he looked in her direction, and she feared that he had caught her. But still, she didn't move as the beating rhythm of the drums kept her mesmerized.

The drumbeat softened a bit, and the older man in the fountain spoke loudly. Cass couldn't quite catch on as it was inaudible; they were far away from her. The older man motioned the three men in robes to enter the pool. The men disrobed and walked into the fountain. The drum beat continued, and the older man said, “Shed your tears in the fountain! Shed your pain, shed your hurt, shed your fears! Cry out but leave one of your tears so you can live your new life without regret and with honor. Remember, only when you shed your last tear will death come.”

The drums beat louder, and the men around the fountain beat their chests with clenched fists in tune with the drums. The three men started to cry and wail.

I can feel their pain, Cass thought, upset.

One by one, the men stopped crying. And as they stopped, they started beating their chests to the same beat as the drum. Cass felt moisture on her cheeks and realized that she was also crying. She peered one last time through the crack and turned away from the door.

Leaning her back on the wall, she bowed her head and dried her tears. After calming herself, she walked back to the painting.

Not a silly story now, she thought.

Chapter 12:

R and D

With Cass and Tor fifty kilometers away, Blight at R and D watches modified arms getting produced. Blight had always been a ruthless leader, willing to do whatever it took to achieve his goals. He was the head of R and D, a secret organization that specialized in creating advanced weapons and technology for the military. They had created many successful weapons in the past, but their latest creation, the heart weapon, had been a failure. It was a weapon that could literally stop a person's heart with the push of a button. But it had been too powerful and too dangerous and was complicated enough to have it completed.

Blight had never been one to give up easily. He had been determined to find a way to make the heart weapon a success. But now, it seemed that his efforts had been in vain. The scientist he had been talking to had just told him that they were unable to recreate the weapon. Blight was furious. He couldn't believe that all his work had been for nothing.

"The weapon's completion requires complex calculations that are impossible to do in such a short time. We need more time to get this done, Mr. Blight," said the lead scientist.

“We have the best of people working on this project. And why is it that we can still not push toward completion!”

“The heart weapon is a dream I have been working on since the start. Only I can envision what this weapon will do. I am a visionary, and only a true visionary like me can understand the need for the heart weapon,” claimed Blight, “The world will indeed bow before me!”

“Yes, Mr. Blight, we are trying our best. I am….” Before the scientist could complete, a mod interrupted by walking past them.

Blight saw something peculiar about the mod. It was with muddled legs and looked as if something unfortunate had happened to it.

Immediately turning toward the mod, Blight asked what had happened.

The mod said, “Mr. Blight, there were four of us when we encountered Cass and Tor. It was a fair fight. We had outnumbered them.”

The mod continued, “Just as we were ready to strike, Cass took advantage of a weapon that he had earlier concealed.”

"The weapon was so powerful that it obliterated the rest of the mods…I was fortunate enough to get out of there fast."

Blight passed a rather disappointing look and said out loud, "This can't be happening! I must be quick."

And with that, he turned toward the scientist and said, "Can you make the weapon right now? Answer me in yes or no."

The scientist, who was taken on unexpectedly, kept looking at Blight. He couldn't give a rational answer since he had already explained to Blight the emerging complexities of completing the weapon.

The scientist failing to give the answer, shrugged, acknowledging that he was not quite sure. Blight, who wasn't ready to accept a no or anything else that contradicted his expectation, struck a blow so quickly that it was merely impossible to see what had happened.

A few seconds later, the scientist started holding his neck as blood was gushing from it. Holding his neck, the scientist fell with his face down on the floor.

After killing the scientist, Blight looked at his hand covered with blood and wiped it off on the mod standing next to him.

Chapter 13:

The Mountain Village and The Train

Back at the inn where Cass and Tor were, it was raining heavily. Both of them could listen to the rain falling on the metal roof of the inn. The little mountain village they were at before was quiet except for the rain.

Of the two, Tor was stretched out on a lounger, trying to gain strength as much as he could. Meanwhile, Cass was quietly eating his breakfast and enjoying the sound of the rain and the calm.

Out of nowhere, Cass picked up clicking sounds coming. Upon focusing, he deduced that the sound was of mods coming into the inn.

Without wasting any time, Cass headed straight to where Tor was resting and tried waking him.

"Tor! Tor! Get up. Someone is coming this way for us."

"What is this sound? It feels like there are mods headed our way," exclaimed Tor.

The two quickly glance out and find the mods walking past them.

Seeing Tor and Cass, one of the mods stop and addresses them both, “We are here to recover the DNA hunter. Whoever has it must cooperate or else be ready to face the consequences.”

The situation, however, somehow felt different as Cass and Tor felt that the mods did not know where to look. Realizing this opportunity to escape, Cass said to the mods, “Well then, you find your DNA hunter.

Cass wished them the best of luck and got back to her coffee. Tor paid for the meal and then walked out of the inn with Cass.

It was only a few seconds later the duo noticed four mods walking across the street and toward their way. The mods had finally found out that the duo was the one they were looking for.

Cass and Tor quickly assessed the situation and decided to split up. She took on two of the mods while Tor handled the other two. The mods were heavily armed and skilled, but Cass and Tor were well-trained and experienced fighters. Cass used her agility and speed to dodge and evade the mods’ attacks, while Tor used his strength and brute force to overpower them.

As the battle raged on, Cass and Tor were able to take out the first two mods with relative ease. The two mods were on the ground with their internals out. Seeing this, the other two mods had geared themselves to their maximum potential. They would not leave without fighting until the end.

Seizing the first move, one of the mods proceeded forward and managed to land a hit on Cass. The blow left Cass flying and, on the ground, injuring her. Whilst the second mod advanced forward to hit Tor.

Tor, who was ready to face the mod, diverted his attention toward Cass and rushed to help her. Realizing how badly Cass was hurt, Tor became enraged. Somehow witnessing the attack fueled a fire in him, leading him to unleash a fierce barrage of attacks on the last two mods. He stashed toward the mod that hit Cass and kicked so hard that his leg went through the mod's body.

The other mod who witnessed this brutality decided to escape. But before the mod could get out of sight, Tor approached from the front and grabbed the mod by the neck. He then held the mod's neck with both of his hands and squeezed it so hard that the neck separated from the body, leaving the internals to come out.

With the fight over, Cass and Tor quickly checked on each other's injuries. Cass's injury was minor, and she was able to walk it off. Tor, on the other hand, was unscathed. The duo took a moment to catch their breath and assessed the damage. The street is littered with the bodies of the mods.

"Well, that was unexpected," Tor said, trying to make light of the situation.

"Yeah, and something tells me it's not over yet," Cass replied as she got off the ground. "There are more mods coming."

"We must hurry before we get into more trouble. Are you able to walk, Cass?" asked Tor.

"Yes, I can for now," replied Cass.

And with that, Cass and Tor quickly gathered their belongings and made their escape, knowing that the fight was far from over. The rain continued to fall as they disappeared into the mountains, ready for whatever came next.

Chapter 14:
The Blast

The rain slowly fell, the big droplets hitting the ground and forming a puddle. Tor knelt both his swords down on the floor and pointed them outward. His face was covered in mud, with blood and oil running down his neck and clothes. Cass was beside him, leaning on him, cautious and alert of the danger around her. The mods surrounded them in a circle, staying out of reach and not approaching. But Cass and Tor knew one wrong move, and they would all lunge at them.

"How are you doing?" Cass spoke quietly to Tor.

Tor breathed in deeply and said, "Just enjoying the rain."

Cass shifted a little, still leaning on Tor, he looked up at the sky. The weather was about to worsen; she needed to do something before the mods attacked them. With one last glance at the mods, she began adding a modification to her weapon. Cass began contemplating what to add to stop or halt the mod enough so the partners could escape.

Suddenly, an idea came to her mind. Cass programmed a big cone weapon that looked like a top pyramid.

Alright, let's do this! She thought and pointed her arm to the ground. Cass instantly hooked her leads to the weapon and grabbed the other leads from Tor's bracelet. The heartbeat monitor started to speed up just then. He breathed in and out, and she matched his breathing. The monitor kept beeping stronger, and Tor gained strength from it.

"I am going to hook up to the harness," she told Tor. "You will have to carry me."

"Okay," he said and nodded. He braced himself and wiped the raindrops running down his face.

She leaned on him more and asked in a hushed voice, "Are you ready?"

He nodded.

"3…." She began to count down. "2…1!"

Swiftly, Cass pulled the trigger of her weapon, which was pointed to the ground, and a pressure blast hit the ground and radiated outward. The blast bubble stopped the rain around them, pushing outward and hitting the mods with the blast, knocking them back.

As soon as the blast hit them, Tor sprang up and ran at full speed toward the train platform. Cass was a dead weight on his back hooked to the harness. While he ran, he could hear the heartbeat monitor still beeping from Cass's heart. He had about 200 meters more to go, but unfortunately, some of the mods started getting up from the blast. They all began attacking the partners, stopping them from escaping.

Tor met the first one with blunt force and just ran him over. Another mod stood up as Tor passed by him. But before he could attack, Tor grabbed his face in the palm of his hand and drove forward. Without missing a bit, he threw the mod forward to the ground.

Chuff.... Chuff....

Cass and Tor's ears perked up when they heard the sound of the train. To their delight, the train was pulling up to the platform.

"Finally!" Cass mumbled. "We are going to make it, Tor!"

"Look after yourself and be safe!" Tor grunted.

"What—"

Before she could ask him, the train's door opened, and Tor unhooked Cass from the harness. In a swift movement, he threw her into the open door. Cass hit the floor with a thud, the impact of the push jolting her.

Tor!

Hurriedly, she looked up through her foggy eyes to see the door closing. Tor stood on the platform with his swords drawn, ready to fight the mods running at him.

No!

Cass tried to get up but stumbled back to the ground. In her haste, she began crawling to the door, but to her horror, the train picked up its pace and started moving. Cass couldn't lose hope; she didn't give up and still crawled. Cass staggered to the door and gasped. She caught the fading glimpse of Tor engaged in combat with the mods recovering from the blast.

Succumbing to her injury, Cass lost consciousness and fell to the floor exhausted.

Chapter 15:
The Alley

The DNA hunter flew through the streets and just below the rooftops from the steam and wires hanging from the building. It was on a faint trail of the DNA, processing what would be considered a faint red footprint and transmitting it back to the mods.

The hunter would need his expelling steam and unneeded particles so that it could track the DNA. Flying and floating through the narrow streets, the hunter slowly tracked the young female, who was its top priority. The hunter slowed down and floated lower. It turned into an alley, moving along with the female.

The sensors on the tracking radar started blinking green, and the shape of the female appeared in front of the hunter. Before it could move, a full-blast percussion wave hit the hunter. The hunter stumbled back and fell to the ground. Another blast hit the hunter before he could recover and rendered it inoperable.

Its hunt was over.

***

From the shadows, Cass walked out and stood over the hunter. She quietly listened, trying to see if there were more hunters around her. And soon, she heard the annoying buzzing noise she had come to hate.

"Damn!" she muttered and looked up.

Another hunter was hovering above her, waiting to attack. She could hear the mods clicking close by. They were moving down the stone streets toward her. Cass stood up and began taking deep breaths, powering up her weapon. In between the beep of the weapon powering, she heard a small bell sound above her. She looked up and saw it was coming from a rooftop.

Cass looked down the alley toward the street and fired her weapon. The first mod turned the corner only to meet a percussion wave coming at him. Cass walked backward, trying to make some space. She could hear the bell ringing louder; this time, it was a little farther down the alley.

She felt it was trying to lead her.

She turned around and ran toward the ringing sound. As soon as she came out of the alley and onto the street, she found a mod running down the street at her. Immediately, she shot the mod and knocked it down. The ringing bell somewhere above her moved in the direction of where the mod turned up from. Cass turned and ran past the mod, kicking him in the face.

The ringing bell now moved down another alley. Cass looked up but could not see where the sound of the bell was coming from. The buzz started again, and she glanced up to see the other hunter flying above her. Quickly, she shot up at it, but the hunter only wavered a little and kept flying.

"Damn!" she said, watching the hunter escape. "Too far away!"

The bell rang again; this time, it was down another alley. With the hiss of the hunter and the clicking sound of mods approaching fast, Cass decided on her next move.

She breathed in and thought, *might as well follow the bells.*

She ran down the alley and paused at the end of it. The hunter was still above her, and she could hear the sound of a single mod walking toward her from the side. Cass crouched down and pulled out her short sword. She waited for the mod to round her. And as soon as the mod crossed the alley's opening, Cass sliced the mod in the knees cutting off one of his legs.

The mod fell, and before he could yell and alert the others, Cass shot him in the chest with her weapon, knocking him out completely. Cass grabbed the mod's separated leg and took the fluids oozing out of it. She began to cover her body with it, trying to disguise and confuse the mods. She leaped out of the alley and ran in the direction of the sound of the ringing bell down the street.

The hunter stayed hoovering in the alley and didn't follow her this time.

Relieved, Cass smiled a little and turned down the alley. Out of nowhere, a rope dropped down the side of the building and in front of Cass. She instantly grabbed hold of it and climbed the 2-story building. She was met with a familiar smiling face when she crested the roof.

"Long time no see, Cass!" Ben smiled at her and waved.

Cass grinned when she saw that the cut on his hand was now healed.

Chapter 16:
Captured

Tor sat quietly in a white room with no windows and just a single door. Right across from him was a chair, the exact replica of the chair he was seated on. Beams of light were radiating from the floor. He could feel several stinging sensations on his back. Tor was sure somebody had tranquilized him, but he wasn't sure who.

He couldn't remember how he got there. All he could recall was Cass leaving on the train while he had a face-off with several mods. One minute, he could be seen putting up a good fight with them, and the very next moment, he felt sharp pangs of aching in his shoulder and back.

Tor sighed and glanced around the room. He was a little dizzy, but overall, he was fine. *Why would they take me?* he thought, grimacing. *What will it benefit them?*

While he was immersed in his thoughts, the door to his left opened slowly, and in walked four men. Tor leaned forward in his chair and observed them closely. *Three mods and one man who seemed to have authority over the others.*

Tor assessed them but remained cautious. He could easily overpower them and burst through the door, but he wanted to see where this 'kidnapping' would go. It was more important than the idea of escaping at this point. Tor watched as the authoritative man sat on the chair across from him. The mods took a position at the man's side and stood behind Tor.

The man stared at Tor and stated, "I am Mr. Blight."

Tor remained silent, staring at him and trying to comprehend his motive. Ever so slowly, he nodded in acknowledgment. Blight gestured to Tor's wrist and said, "That is a nice bracelet."

Tor persisted in silence, never uttering a word. Suddenly, the mod to Tor's right grabbed his arm and tried to get hold of the bracelet. Enraged, Tor pulled his arm away, and before anyone could blink, he grasped the mod's throat.

Tor pulled the mod closer to his face and snarled, "It is a gift!" He glared at him darkly. "I don't think you can have it."

He pushed the mod away, making him stumble back. The other two mods roared and reached for Tor. Blight was watching them with amusement and curiosity. He waved his hand, indicating the mods to stop. The two mods nodded and hobbled back to their positions. Blight looked at the mod that grabbed Tor and pointed his hand at him. A spark ignited from his hand, sending a percussion wave in the mod's direction and knocking him back against the wall.

"No one said to touch him," Blight informed, staring sternly at the mods. "A teardrop is powerful." He looked at Tor sideways. "He can discard you all anytime he wants."

Tor rolled his eyes at Blight's exaggeration and leaned back in the chair, untouched by the situation.

Blight gave a predatory smile to Tor. "I like the fascinating little weapon system your friend has. I find it really interesting and intriguing. I would like to have it within my hold."

A dark glint projected through Tor's eyes, but he didn't say anything.

"Don't worry; we won't harm you as you finally understand why you are here. No? Oh, come on, Teardrop. I thought you were better than this." Blight let out an exasperated sigh. "You are a bait, nothing more." He chuckled. "How does it feel to be used?"

Tor half smiled and said, "I guess I will have a nap and wait."

One of the mods said, "We should torture him and take the bracelet from him."

Blight raised his hand again and remarked, "Torture will do nothing; it is a waste of time. I will get his last teardrop later."

Grinning, Tor leaned forward and beheld Blight in the eyes. "Don't hurry to meet death; there is a place for you."

"For us all, Teardrop; for us all," Blight riposted. He stood up and glanced around nonchalantly. "Make him comfortable. Any harm to him and you will all pay; if...you happen to survive his wrath." He chortled, turning around and walking away. "Shed your tears so that you can live with honor."

"There is no life without honor. I will not shed tears," Tor grunted. "I will wait for one day before I decide to leave."

Blight turned. "I will look forward to the opportunity." He slammed the door behind him, leaving behind an echo.

Tor sat back and glanced at the mods. Grinning, he said, "Get me something to eat; I am hungry."

Chapter 17:
The Scavengers

Cass followed Ben across the rooftop. They crossed several roofs and then came to the industrial market side of town. They climbed up and down a series of ladders and scaled a few ledges. Eventually, they both arrived on top of a large manufacturing building. The building was at least 20 floors high and vacant, but there were countless shacks and fires on top. Cass saw a small market and almost a town on top of the building and the surrounding buildings.

She followed Ben closely as he weaved in and out of the people and the buildings. Ever so often, he would turn and urge her to keep up. Cass humored him and acted like she was trying to keep up. And soon, the duo entered a small building that was part of the plant and built on it. Ben led Cass to the side where there was a room.

It's a big office, Cass wondered, roaming her eyes around the room.

Sitting behind a desk with a stack of papers was a brawny woman with curly hair and a pleasant smile that seemed to be on her face all the time. She was reading and sorting documents while the others in the office ran errands. It seemed like they knew what to do with every flick of the lady's hand.

Cass smiled and stood at the door. The woman glanced up when she felt a movement and waved Cass in, saying, "Have a seat." Cass started to introduce herself, but the lady interrupted her. "I know who you are." Cass raised an eyebrow in question, to which the lady chuckled. "You can call me Ms. S." Cass could see that even though the woman was all about business, she was cheerful and had the time to laugh.

"You and that teardrop have been creating much buzz around," she stated and flipped over a notebook. "It seems like Blight really wants your weapon." Ms. S paused. "He has even captured the teardrop."

Cass looked at her bewildered, about to stand and run out of the building. *How did she know all this?* she thought. *Just who the hell is she?* She had to get Tor, but before she could move, Ms. S glanced at her sternly as if stopping her.

After a minute, she asked, "How do you know this?"

Ms. S. smiled and said. "My scavengers know everything; they pick up the pieces after every fight. We recycle the mod parts and gather endless information. So we know all that goes on." Cass looked around; she could see the hustle and bustle of the office with people coming in and out.

Ms. S looked at Cass through the top of her glasses and cheerfully said, "Before you run off, let us look at this weapon that can blow trees down."

"I think I need to go find Tor," Cass said, smiling at her.

"We know where he is."

"What?"

"He is being used as bait to lure you to give up the weapon."

"I need to tell Mom," Cass commented.

"That hussy," Ms. S giggled. "It will be better if you stay with us for a little while. Then we will help you get the teardrop back, but with a fee, of course." She stood up from the desk and scurried around. "Come on, child. Let's go."

Ms. S moved off briskly. The woman's speed surprised Cass; she was fascinated by how the elderly lady walked. Cass had no choice but to follow her, and Ben walked beside her. Soon they were all outside again, walking to another building. As Ms. S moved, people seemed to move with more purpose. Cass could tell the woman was in charge.

Out of nowhere, Cass heard the mod's clicking noise and instantly reached for her weapon. She looked around and saw a cloaked figure standing in the shadows just within the peripheral vision of Ms. S. Ben noticed Cass grabbing her weapon and looking in the direction she was staring.

Ben grabbed her hand, stopped her, and said, "Wait! He is with us." Cass stared at the little boy as if he had gone mad. "That's Frank. He is Ms. S's bodyguard."

Cass stared at him walk out of the shadows. Her eyebrows moved up in surprise, gazing at the stranger up and down. Frank had a patchwork of mod parts; arms and both legs were modded, but his legs were not the same length. Even though he limped, he still moved quickly. Cass slowly placed her weapon back in its holster, and the beeping sound of her heart stopped at once.

"Fascinating!"

Cass heard a bland voice mutter under his breath. She turned around to see Ms. S standing beside a slender man with a mod arm.

"Come on, let us look at you," the man said dryly.

Chapter 18:
The Science

"Cass, can I see your weapon," Ms. Sim asked Cass.

Cass contemplated at first and then sighed and handed the weapon to her. Ms. Sim looked it over and passed it to the scientist. The scientist examined it and mumbled under his breath. When he was done, he looked over at Cass to say, "It is just a hand-held pressure weapon powered by the pressure from the heart."

"I have a strong heart," Cass smiled and ensured.

The scientist looked at her cautiously and then handed the weapon over. He pointed to a place at the corner and said, "Can you shoot those targets?"

"What mode do you want me to shoot with?" Cass asked. "I have a rifle, both normal and area blast."

"Normal is fine," the scientist said.

Cass connected the weapon to the bracelet, and the beep started automatically. She fired her weapon at all five targets, knocking them over. The beeping noise of her heart monitor stayed steady the whole time.

Cass smiled and told them, “When I am linked to Tor, I blast big things such as trees.” She smiled at Ben. “Or goats.” Ben grinned back at her

The scientist stared at her, his gaze firm. “Your heart isn’t that strong.”

Pointing at her chest, Cass smiled and said, “Yes, it is.”

The scientist walked closer to her and outstretched his hand to touch her chest. Not understanding his intention, Cass quickly drew her knife and put it next to his cheek.

“What the hell are you doing?” she said flatly.

Ms. Sim laughed out loud and said, “Oh, child, you can see he has not been around many women.” She winked at Cass and waved her off. “Put the knife down.”

Cass grunted and pulled her knife down. Confused, the scientist’s gaze darted between the two women, making Ms. Sim and Cass chuckle. He huffed and rolled his eyes, walking away. Before he could vanish through the corner, he waved his mod arm and said, “Alright, enough of it. Can you please walk behind the x-ray?”

Chapter 19:

Dinner

Cass sat with the little boy on a ragtag cushion with a plate of food they were sharing. Ms. Sim was back to work ushering people this way and that. Everyone could hear her joyous laughter around. It seemed to keep people happy and working.

"Where is your mom?" Cass asked Ben.

Ben turned to look at her and said, "I don't know, really. She left with a lady and hasn't been back yet."

Cass looked sideways at him and asked, "She left with a lady?"

"Yes!" Ben said. "We got caught by some people when we were in town. There was a pretty lady with a hood on her head. She spoke to my mom, and my mom said she would be away with her for a while."

Cass smiled at him, realizing his mother would pay her husband's debt. She didn't know how and what, but she felt bad knowing it would leave Ben without his parents.

"What did the women look like?" she asked.

"She was tall and pretty," he said, smiling at her. "She held my face in her hand, kissed me on the cheek, and told me I should always remember this kiss as a debt paid or something like that." He shrugged. "Then she dropped me off on the rooftop over there," he pointed in a general direction. "From then on, Ms. Sim has been taking care of me till my mom comes back."

Cass got up and told Ben that she would be back. Once outside the room, she caught sight of Ms. Sim and approached her.

"How many do you take in?" she asked.

Ms. Sim smiled and said, "I take all of them in." she paused and sighed. "Anyways, how was dinner? I hope you enjoyed it."

Cass stood there wondering where Ben's mom was. She swam out of the depth of her thoughts when she heard the clicking sound of a mod coming up. She turned around to see Frank walking toward her. Under his cloak, she could see the misshaped legs and arms. Frank stopped beside her, looking in the same direction she was.

"When I lost my legs," he spoke to her softly, "I sat down there," he pointed to the street below. "I could never afford mod legs, and then I lost my arm when a person cut it off to replace his arm. They were supposed to give up. But ended up taking mine." He sighed. "I laid there, waiting for my death when a hooded lady appeared and carried me up here." He glanced at her. "Ms. Sim took me in years ago. I received these mods from parts scavenged from the streets. Now I protect this place."

Cass looked at him and smiled, nodding.

Frank smiled back and said, "I am afraid your heart will destroy this rooftop city when they come for it. You need to leave."

Cass's smile turned to a frown, and she stared at him. "Don't worry, Frank; as soon as I know where Tor is, I will leave," she told him.

Frank spun at Cass, swinging at her, throwing his cloak, and the full misshapen body and parts were visible. Cass jumped back, dodging the swing. She was about to grab her weapon but decided not to when she saw Frank moving quickly at her. Cass dodged his kicks and punches, blocking blows.

"Stop!" Cass shouted.

"I must keep this place safe!" Frank retorted, attacking her.

Cass kicked the smaller leg from under him, and he fell to all four. Quickly, he got back up and charged at her. She noticed all of his limbs were mod. Cass dodged the charge, spun to the left, and pulled out her weapon. She jumped on Frank's back and used a small blast to shoot his head, knocking him out.

Frank fell to the ground. Cass looked around and saw Ms. Sim looking at her. She was about to speak when Ms. Sim raised her hand and said, "I know that Frank can be overprotective. Let him sleep it off. We will go finish dinner."

Chapter 20:

Sunrise

The next day, Cass woke up and stretched her arms, yawning. She peered around and noticed Ben sleeping beside her, snuggling. She smiled lightly and looked out the smeared window to the first light coming through the mountains to the east.

Cass sighed and quietly walked outside, looking across the rooftop city. She noticed smoke in the air and realized it was coming from the kitchen's direction. Everyone was getting ready for the day. Steam rose from the buildings below, and the sun shot light beams, casting shadows between power poles wires. Occasionally, a rainbow from behind a steam plume would appear as if playing hide and seek.

Cass took in a deep breath and let it out.

"Pretty, huh?"

She heard a small voice from behind her and turned around. Ben had woken up, and his tired eyes rotated from her to the sunset.

Cass smiled at him and said, "Come here. Let us watch the sunrise together."

Ben nodded.

Cass and Ben stood side-by-side for the next few minutes and looked at the sunrise. Suddenly, Ben pointed below the shadows and said, “Look at that.”

Cass watched the steam and light coming from the east. She squinted her eyes in the sun and looked closer.

She was about to say something when suddenly an alarm started going off in the distance. Surprised, she watched as the shadows revolved into more than a dozen mods. They were heading in the building’s direction but were moving slowly.

That’s odd! Cass thought, frowning. *Why are they—?*

Her eyes widened when she realized the reason as one mod became visible. Hanging under the hunters were mods being flown to the rooftop.

Cass looked at Ben and said, “Run and hide.”

When she was sure Ben would be safe, she drew her weapon and started running to the east side of the roof. She saw the scavengers running and heard the click of mod parts running past her.

Frank passed her, snarling, “Do you see the destruction you caused?”

Cass said nothing and raced beside him.

The first hunter dropped the first mod off. Before the mod could get up, Frank tackled him and ripped the mod's arm off. Frank stood on all fours like a spider, switching left and right. Cass shot, hitting a hunter, making it drop the mod to the ground below. She aimed and shot again but missed it as the hunter dodged it. She looked to her right and left and saw other mods being dropped off on the rooftop and hunters flying back down to the ground.

Cass looked down the building and saw mods climbing up the wall. She heard a shout and spun around to see several mods jumping on Frank. Instantly, she ran toward him and blasted two of them. Cass pulled her short sword out; she slashed with one hand, and she shot at the mod parts with the other.

The number of mods kept increasing on the roof. Cass saw Ms. Sim with a mace in her hand, smashing it into a mod's head. After dodging all the mods, Frank yelled at Cass, "Get back to the center of the roof!"

Ms. Sim waved all the people back as she struck another mod with her mace. "Come!" she shouted at Cass and Frank.

Frank and Cass sprinted to where Ms. Sim was. While Frank jumped over several stacked tires, Ms. Sim ordered everyone around her. In the midst of it, a massive breaker got switched on. The whole building shook as a large electrical surge coursed through the building.

The vibration and pressure were so strong that the mods started to fall on the floor. Hunters that were close were shocked as well. Frank grabbed Cass by her arm and lifted her slightly off the ground. Cass felt a little tingle in her arms and heart.

"Might as well save you from the destruction," Frank grimaced at her. Cass nodded and hid her face in his chest.

***

Fifteen minutes later, the electric surge was over, and the rumble stopped. Ms. Sim's workers on the rooftop city acted quickly, taking the mod parts and throwing the bodies down a Shute to the ground below.

When Cass saw them, her eyes widened, and her mouth dropped.

"That is the reason we don't get bothered," Ms. Sim said, walking toward Cass. "It is a messy and hard way to live. But most up here would be chopped to pieces and body parts sold. It is the lesser of evils, I always tell myself."

Cass looked down at the ground and softly said, "I understand."

She looked around for Ben, only to find him picking up mod parts. She turned toward Frank and said, "Thank you for saving me."

Frank nodded but didn't say anything. Cass sighed and glanced back to the east. The sun was up, but there were no shadows, and only the rainbows were in view now.

Chapter 21:

Plans

After discarding the mod's part, everyone left except Cass and Ben. They stood near the edge of the building, looking across the rooftop. Unexpectedly, she felt someone move on the rooftop adjacent to her.

Who are they? She thought, narrowing her eyes.

The figures moved and looked in Cass's direction. There were two people. One was in a hood and had a silhouette of a tall woman, while the other was an average size man. Cass watched as the man panted and sniffed and kept touching his face.

It's the sniffing man Tor and I met earlier!

Cass stared more intently as the light grew. The man looked into her eyes and smiled. He sniffed and waved at her and Ben, then turned and walked away. The women also turned and followed him, moving effortlessly. The hood and the flowing cloak obscured her; it made Cass difficult to recognize her.

Cass looked down at Ben and said, "Can you do me a favor?"

"Yes, sure!" Ben said, smiling.

"Can you follow them for me?" She tilted her head in their direction. "They won't notice you."

"No problem!" Ben looked in their direction, grinning. "What are you going to do?"

"I think I am going to talk to someone important," Cass smirked. "You be safe. I will see you back here this evening."

"Yes, I will see you here." Ben puffed up his chest.

Cass watched the two that were on the roof disappear out of sight. After she saw Ben quietly following them, Cass turned and walked to the ladder that led her down the roof.

***

The Blight company had a huge office building with a sterile-looking industrial frontage. Cass stood at the door and sighed. *Here we go!* She pushed a button and waited. A second later, she heard a voice asking her to identify herself. Cass told her name and that she wanted to see Mr. Blight. While she stood there waiting at the door, a DNA hunter flew around the building and hovered above her.

A few minutes later, a woman walked to the door and opened it. She curtly told Cass to follow her. Cass nodded and followed her. After going through various corridors, the woman led Cass to a neat and clean office. The receptionist asked to sit down while Mr. Blight arrived. Cass sighed and sat down in a chair. The receptionist left, and as soon as one door closed, another opened, and Mr. Blight walked in.

"How are you, Ms. Cass?" he asked, smiling. His sharp features got highlighted by his smile. He sat down, not at his desk but at a chair beside her. "Now that you are here, I would like to apologize for sending the hunters after you. I didn't think you would answer a formal invitation."

Cass flipped her head, making her ponytail fly, and glared at him. Giving him a sarcastic smile, she said, "Inviting me to lunch would be a lot less expensive than replacing the hunters and mod parts."

"Oh, come on. It's just money." He shrugged. "Every dime spent is worth it, so I can see your weapon."

"Too bad I left it somewhere else." She nodded.

He stood up, walked to the other side of the desk, and sat down. "Cass, I would offer you all the money you need. You could live your life in luxury."

She raised her hand, stopping him, and shook her head. "I don't need money. I want you to leave me alone."

He leaned back in his chair and said, "I would gladly pay you. And let me assure you—one way or another—I will have your weapon."

She smiled and leaned forward. "Not while I am alive."

Cass pulled her knife out and lunged over the desk, landing on Blight and knocking him out of the chair. She pulled the knife at his throat and glared into his eyes. Blight didn't resist or move. He just smiled at Cass. She suddenly felt heat on the side of her face and realized he had the blast weapon in his hand.

"Such reckless behavior," Blight said, running his free hand on the side of her face.

"You have been warned!" Cass said, glaring. "Next time, I won't stop!"

Blight was about to speak when the receptionist quickly ran inside the room.

"Sir!" she exclaimed. "He…He has escaped."

Instantly, Blight shoved Cass off of him. She recovered immediately and stood defiantly staring at him. Blight adjusted his jacket and looked at Cass without care.

"Till next time, little one," he said, walking toward the door. He turned to the receptionist and said, "Make sure she is escorted out safely. I don't want any harm to come to her."

Chapter 22:
A Long Day

Tor crouched in some bushes as two of the four-legged mods walked by. As soon as they passed, he stepped out and ran quickly in the opposite direction. Instantly, he felt a percussion wave hit the limb above him on the tree. The splinters rained down on him, cutting the side of his forehead. Tor continued to run on the path. Several more blasts and limbs and branches flew in every direction.

He turned around the corner, ready to attack the mods shooting at him with the axe in hand. *Not the best weapon,* he thought. He looked down the path and did not see a mod. He was about to move when a blast hit him. Alerted, Tor saw a DNA hunter with a percussion weapon.

"Of course, they would have his DNA!" He grunted and looked around him.

He heard the hiss of several more hunters. Tor scanned the area; 5 hunters flew from the manufacturing building. He quickly ran back toward the building and into an unsuspecting mod. He knocked him to the ground.

Tor felt another blast from a hunter hitting the tree beside him. He grunted in pain when a part of the blast struck him on his left shoulder and caused him to stumble.

Another blast hit close to the ground knocking him off his feet. He rolled to the right as he fell and ran around a tree. This time a blast hit him in the chest, and he took his breath for a second. He noticed the mod that shot him was too far away for the blast to be effective.

Before Tor could move, he was hit from behind by another blast. This one knocked him to the ground. He was on his hands and knees, breathing heavily. Tor looked toward the mod that shot him while the hunters hissed above him.

1....

2....

3!

Tor breathed in deeply and sprang at the mod when he moved closer. He cleaved off the mod's right arm using the fire axe and threw him away. He started running while the DNA hunters flew after him. The blasts from them were quick but not as powerful. But when a blast hit him, it would make Tor stumble.

The hunters were coming at him from all directions, blasting and making debris fly. Some would cut his skin; other blasts would make him stumble.

It dawned on Tor that he was being herded.

Almost too late! He thought.

Ahead of him, he could see the flash of a four-legged mod. Tor stopped quickly. The hunters flew past him and halted. It turned, but by then, Tor had already run to his right. There was some thicket that he wanted to get to. Making it to the thick undergrowth, Tor crawled quickly into it. Crawling through the underbrush, the hunters kept blasting away above him. The blast would not hit him directly, but the thorns and limbs would. Tor continued to crawl, making as much pace as possible.

Suddenly, he felt a brush push and something on top of him. He looked at both sides and saw the four legs of the mod straddling him. The blast from the weapon hit on top of him. It smashed him to the ground.

Tor breathed in deeply, catching his breath again, and crawled forward. He could hear the mod walking but got tangled in the brush. He flipped to his side and struck at one of the legs of the mod with the axe, knocking it off balance.

I need to dodge them!

There should be a way!

Tor stood in the brush and jumped on the 4-legged mod, driving the axe's handle into the mod's chest where the heart was. The mod stumbled and fell into the thicket. Tor used the mod as a scapegoat to jump out of the thicket and back to the way he started crawling from.

The DNA hunters waiting for him started shooting blasts at him. Debris continued to fly, and the percussions would hit him, making him stumble.

Tor ran back to the trees—to a small tall sapling. Using the axe, he swung twice, cutting the tree down. He grabbed the small tree and swung it at the closest hunter. The limbs and branches knocked it off track. The hunter wobbled lower to the ground. Springing into action, Tor struck the hunter with the axe. The hunter fell to the ground.

Tor turned and ran farther into the woods, carrying the axe and the sapling. He noticed the hunters had slowed following him due to the trees. He was using the cover to stifle their movement. He watched the mods waiting outside the forest, hesitant to follow him. Tor wanted to return to the plant, hoping the machinery and closed areas would help. Also, he could find better weapons. When he struck the hunter with the axe, it cracked the handle.

A while later, Tor stopped for a minute using the axe; he cleaned off the limbs and made a spear of one end of the sapling. He threw the broken handle of the axe to the side.

Tor removed his belt and secured it around the axe top. He held the axe from the belt side, getting ready for the mods.

Tor realized that the plant was some 200 meters away now. In front of him were four hunters and ten mods. An opportunity for a great day. With the crude spear in one hand and an axe head tied to a belt, Tor ran for the plant again.

He took two big, deep breaths and planted his feet firmly on the ground. Slowly, he leaned forward and started sprinting toward the mods.

Tor moved fast and lowed, thrusting the spear in front of him. The spear stabbed through one of the four-legged mod's legs. Tor broke the spear off, so the wood was still in the leg, making it inoperable. Quickly, he slung the axe head again, making the belt strike the mod on the side, sticking into its body. Tor ripped it away and jumped to the side.

Out of nowhere, Tor felt a pressure wave bounce off his body, making him stumble. He regained his feet and lunged towards another mod but was hit in the side again by another blast. This one knocked him to the ground. Tor rolled back to his feet, flung the axe head out, and struck another mod with it.

Tor stumbled forward and was hit with another percussion blast. He stumbled again, kneeling on the floor, and felt another wave's impact.

Tor breathed in deeply and focused his attention on the mod before him. He lunged forward, striking the mod. He grabbed him by his arms and spun him toward the direction that the blast was coming from. The mod took the hit but still knocked Tor around, stumbling.

Tor fell to the ground with the mod beside him. He struggled to his feet; his vision became blurry from the impact. Tor could feel the heartbeat increasing; he stumbled and breathed deeply, trying to focus on the next mod.

Before he could lung, a hooded figure appeared in front of him. The cloaked figure pulled out short knives from the cloak. She floated in the air as it moved gracefully and sliced through the remaining mods while Tor gathered his bearings. He grabbed a mod that was nearby, flung him to the ground, and stomped on his chest, pulling the arm from its socket.

The hooded figure ran in Tor's direction and past him. She whispered, "To the woods!"

Promptly, Tor turned and ran with the figure. They both ran, dodging the trees and bushes. The figure moved as if flying while Tor kept up with her. Tor didn't hear anyone behind them and sighed in relief. The figure led them out of the woods and to an open courtyard that seemed to be in a park.

The figure stopped and turned to him. Tor paused a few steps away from the hooded figure and stood facing the figure. The hood was pulled back, and under the hood was the familiar sultry face of Mom.

"Did you see teardrop?" she commented, smiling at Tor. "You needed me, after all."

Tor stared at her as she moved toward him. Mom seemed to float as always as she moved close to him. He put his hand out to stop her.

"Tell me what is going on?" he growled.

She smiled and stopped. "Do you really want to know, teardrop?"

"Yes." He nodded.

"Teardrop," she said, smiling and moving closer. "A few years ago, you were sent on a mission. You were sent to get a prototype weapon and capture a scientist."

Looking at Mom skeptically, Tor said, "I don't remember such a thing."

She moved toward him, but Tor side-stepped her. Mom rolled her eyes and continued, "Well, anyway, you don't remember, but others do. During the mission, you went to the scientist and were about to capture him when the scientist's daughter tried to stop you. You stabbed her heart and threw half of it away."

Tor side-stepped again as Mom circled him. He shook his head and said, "No such thing happened!"

She laughed and continued, "The scientist used the weapon and knocked you out. He pulled this short sword out of his daughter's heart." She threw the sword at him, and he caught it in midair. Tor gazed at the markings of a teardrop sword. "That scientist put the weapon in his daughter, a half-mechanical heart that helps her power the weapon. And the other half of the heart, he placed it inside of you." She paused and looked him in the eye. "The complete weapon system is you and Cass."

I thought the scar on my chest was just another scar! he thought.

Tor looked up to see Mom right in front. She touched his face and said, "Now I know why you and she were dropped in business a few years ago." She smiled slyly. "Now you know the sad story."

"Does Cass know?" he asked, glaring at her.

"Who knows!" She shrugged. "But time will tell."

He grabbed her wrist and pulled her hand from his face. "She will never know! I will carry this burden for the rest of my life!" Tor leaned down and looked into her eyes. "Do you understand?"

"Oh, Teardrop!" Mom smiled, kissed him, and started to float away. "We shall see."

Tor watched as she moved away. *Why does it always look like she floats when she moves,* he thought. He could hear the mods coming in his direction. Tor sighed and held his short sword in his right hand and the axe head in the left.

"Took two deep breathes," Tor said to himself as he turned to the mods. "Death waits for you!"

Tor stood in the small clearing facing the mods and hunter, waiting.

A minute later, four mods came out of the tree line and slowly walked closer to Tor. A DNA hunter that was quickly modified appeared over the trees. Tor could hear the mod's parts clicking and DNA hunter hissing. Aside from the wind and the moving parts, there was no other sound but his heartbeat and steady breathing.

The mods and the hunter just paused out of reach. Tor was preparing for their attack or for him to attack them.

He took in a deep breath and let it out. He was about to take another deep breath when Blight walked into the clearing.

Blight raised his hands to Tor and said, "Just hold on!" He waved his hands. "You are a man of his word; you gave me one day."

Tor tilted his head and half grinned. Blight walked closer and stopped a few steps away from Tor.

Blight was about to talk, but Tor raised his sword and pointed at Blight. He said, "There is no power in this world that can save you if you continue on this path."

Tor dropped his arm, turned, and walked away, leaving them all standing in the clearing.

Chapter 23:

Passage of Arms

Cass walked beside the receptionist, scanning the area. She gauged the movement of people around her. The clicking sound of the mods resonated in the surrounding. Cass could hear the rhythmic click of the receptionist mod arm as they both walked. All of a sudden, she realized that the receptionist was guiding her in a different way. And instantly, heard the clicking of a mod behind her.

Shit! she thought.

The receptionist turned, grabbing her. Cass quickly drew out her knife and slashed the receptionist. The receptionist turned, dodging it but keeping firm ahold of Cass's arm. Cass pushed the lady, smashing her into the wall. The receptionist still kept the grip with her mod arm. To make her lose the grip, Cass stabbed through the mod's arm, making the receptionist let go.

Another mod behind Cass grabbed her, lifting her off the ground. Cass used both legs and kicked the receptionist against the wall, making the mod stumble and fall to the ground with Cass on top. She wiggled herself free and to her feet, stabbed into the mod's leg with her knife cutting the hydraulics of the leg. The fluid gushed out from the hydraulics, covering her and the receptionist.

The receptionist quickly slid down the wall away from Cass and entered an office. Cass quickly followed her, leaving the mod man that had grabbed her on the floor.

Cass busted into the office after the receptionist and noticed on her desk a bundle with a sword and a small axe. She grabbed the axe and went to the receptionist.

"Where did you get this?" she asked, furious.

The receptionist said nothing and charged Cass. She side-stepped and grabbed the receptionist by the cuff of her shirt; pulling her back, Cass slammed her on the desk. She swiftly cut the mod arm off.

"Where is he? she growled, trying to come up with her best 'Tor' impersonation.

"He has escaped!" the receptionist cried. "It's the reason why everyone is in a frenzy!"

Cass shoved the girl off the desk and grabbed the bag with Tor's stuff in it. Holding the axe in one hand and the bag over her shoulder, she quickly ran out of the office. Cass turned and started sprinting to the front door, and as she did, a pressure blast hit the bag, making her stumble. Cass didn't give up and crouched forward, running harder.

When she was near the exit door, she saw a 4-legged mod standing in the way. Cass ran hard, sliding between his legs and using the axe to cut through one of the legs. She hit the door with her feet and pushed it open, springing to her feet.

***

Cass ran outside the building, knowing the mods and DNA were chasing her. She ran past and continued to a wall that was outside of the building. She quickly climbed on top of it. Once on top, she picked up her bag and peeked inside it. Inside, it was a pressure weapon, already in rifle mode. Cass connected the weapon to her, and the rifle pressure sprang to life from the rapid beat of her heart.

She lay in the prone position on the wall and, taking aim at the mod coming out the door; she shot, knocking the mod down. The receptionist ran out of the building with mods following her.

They never learn, she thought, chuckling.

Cass shot the receptionist in the leg, knocking her to the ground. She then rolled to the other side of the wall and ran off a little farther. Ben was waiting in the alley of the street. She ran to him and handed him Tor's bag.

"You give Tor this and tell him to go to the service company; I will meet him there," she instructed him.

Ben nodded and ran off down the alley while Cass ran in the other direction, down the street. She saw a DNA hunter flying over from the building. She knew it would find her soon. Cass ran quickly in the other direction and reached an alley. She turned and grabbed a fire escape ladder, climbing to the rooftop.

Once at the top, Cass paused and pulled her knife into her braid with the tip sticking out. She quickly flung the braid over her shoulder and looked down the alley. The hunter had just turned around the corner with several mods moving quickly behind it. Cass took the modified rifle in her hand and held the pistol and the axe in the other.

The DNA hunter started to gain elevation. As it did, Cass leaped from the roof and landed on top of it. She took her pistol and shot the hunter causing it to spew steam and fuel it. It started to descend, and Cass rode it down to the ground shooting it one more time.

The mods were stunned when they entered the alley, and taking the upper hand, she shot a pressure wave at one of them. The wave knocked him down, making Cass run in the middle of them, shooting at each of them and swinging the small battle axe.

She would spin with each swing and shot and exaggerate the swing of her head. Each time catching them off guard with a slice from the knife in her braid. With each spin, there was a blast, a kick, a strike with the axe, and then a cut from the blade.

As soon as there was an opening in the alley, Cass ran toward it. She turned around and looked back at all the mods lying and sitting, tending to their wounds. She smirked and saluted with the axe. Turning around again, Cass slung her head and caught her braid with her hand, and ran as the rain started to fall down.

Good, she thought. *The rain masks DNA*. She had to make it across town to the broken hearts service company before the enemy got her hands on her.

Chapter 24:
Aftermath

Blight stood in his office looking out the window to the woods he had just come from. He was aimlessly swirling a glass in his hand, thinking how the mods could only produce one good blast with the weapon, but Tor seemed to have endless energy. Soon, he could hear the faint alarms in the building and sighed.

"Can't they do one thing correctly?" he grumbled. "What am I going to do with them?"

The door interjoined with his secretary's opened. Blight turned around and saw Ms. Linda walking with a limp and immobile arm. He sighed and sat his drink down. He walked to her and smiled.

"Ms. Linda, please sit," he told her and escorted her to a seat, taking the paperwork from her hand. He walked over and retrieved a mug filled with water.

"Ms. Linda." Blight smiled. "Why did you do that?"

Ms. Linda smiled back and said, "I got caught in the moment."

He smiled at her again and asked, "Where are the others that were with you?"

She sipped from the glass and said, “They are standing outside.”

Blight walked to the door and opened it, looking at the two mods standing there. He grabbed the mod closest to him, pulling him close. “Your purpose was to detain the girl and ensure Ms. Linda is safe,” he snarled at the mod. “You failed both.”

With his free hand, Blight grabbed the mod’s arm and blew the arm off using the pulse in his hand. He let go of the mod and grabbed the arm that fell to the floor. With the destroyed part, he started beating the mod with it. The fluid from the arm and the blood from the mod man slung around with each blow. Blight continued to kick and punch the man. While the second mod man stood there watching in horror.

Blight turned to the second one and shouted, “You failed as well!” he grabbed him. “Ms. Linda was injured because of you.”

Blight lashed out at the mod, pushing him to the ground. He was breathing heavily and snarling rabidly. He took his hand and covered the other mod's face. Sitting on top of the mod, Blight blasted his face with the pulse wave and then shot the mod in the face with his other hand. The second blast disintegrated the mod’s head on the floor.

H stood and straightened his clothes, returning to his office. Ms. Linda was still sitting in the chair, sipping from the mug. She looked up at him and, smiling, said, “Clean yourself up.”

Blight turned, grabbed a towel from his chair, and wiped his face. Ms. Linda stood up and walked toward him; she took the towel from him and slowly cleaned his face.

Once she was sure he was cleaned, she straightened his collar and said, “You do too much.”

Blight looked down at her, staring into her eyes, and said, “Never too much for you, Ms. Linda.”

He moved back from her and went back to his desk. Ms. Linda sat down again in the chair across from him. Blight hit the intercom button and said, “Send someone from the science department in.”

“I am right here,” Ms. Linda said, chuckling. “I will get them for you.”

“Okay.” Blight laughed. “But make sure they fix your arm and have your leg looked at.”

“Look over the paperwork on your desk.” She smiled at him. “You will find something interesting.” He winked at her, and she smiled back at him. “I will get the mess cleaned up outside.”

He nodded and said, "Also, can you please ask them to turn off the alarm."

Ms. Linda nodded and walked out. He could hear her already starting to give orders to the passersby. He watched as they all quickly ran away to do the task as told. The door closed as Blight picked up the papers on his desk and started to read them.

***

Twenty minutes later, Blight looked up from his desk as Ms. Linda escorted two scientists in. He motioned for them to sit and smiled at Ms. Linda. He could see she had not tended to her injuries yet. He was about to say something when she just turned around and walked out.

As soon as the door squeezed shut, Blight looked down at the papers on his desk. He tapped his fingers on the desk and, without looking up, asked, "Well, who wants to start?"

The scientist looked at each other and gulped, shuffling in their seat. When none of them replied, Blight got irritated and looked up. He nodded to the one on the left and said, "You start."

The man paused and took a breath.

"Sir," he fumbled, "When the specimen was in the cell, we monitored how his heart and lungs functioned using the sensors. We found that the specimen has an abnormality."

Blight raised his hand, stopping him, and grunted, "I can read that in the paperwork." He flung it at the man. "What is the abnormality?"

The other scientist spoke—rather quickly, "The specimen has an extra valve on his heart."

"The specimen is a man!" Blight exclaimed, frowning at them.

"Yes, that is true," the scientist answered. "It seems that an extra half part of a heart was surgically affixed to his heart. From the sensors, we found out that before he smashed through the door, he coughed and took three deep breaths causing more blood to flow to his muscles and lungs, giving him explosive power."

Blight stood up from his desk and walked to the window. He sipped his drink and turned back to them, "Go back to work; I want you to work on the mechanical heart." He stared at them intensely. "It is still not used in the battle."

Chapter 25:
Ambush

Back in the forest, Tor moved swiftly through the trees. He had to get to the service company! It had started to rain, so it masked his movements a little bit. He could not hear anyone chasing after him.

As soon as Tor crested out of the trees and into the park, he could see a small figure carrying a big bag walking across a bridge. He slowed his movements and stepped closer. When he was closer, Tor realized the silhouette belonged to Ben.

"Ben!" Tor called him. "What are you doing here?"

Ben paused and turned around. Tor observed the strain on the boy's face, which quickly turned into a smile when he saw Tor.

"I have your stuff!" Ben exclaimed. "Cass gave it to me and told me to find you."

Tor walked up to him and picked the bag off his shoulder. He opened it to see his weapons inside it. Ben smiled when he saw the weapon in the bag. Tor grinned, observing his reaction, and pushed a small knife toward him.

"For whenever you are in trouble, kid." He smiled. "Now, get out of the rain."

Ben looked at Tor and then at the knife, contemplating it. Swiftly, Ben stood a little taller and ran across the bridge. Tor watched from behind his hooded cloak as the boy ran across the bridge. He shook his head and muttered, "The simplicity of a child."

Tor collected his gear and stood up. He looked at the sky and surveyed the area. The rain was coming down a little harder now, and the wind was picking up. This was an excellent time to move. The DNA hunters had a hard time in this kind of weather. Tor took off in a jog in the opposite direction the boy went into. He wanted to buy time to get to the service company to meet Cass.

***

Two hours later, Tor arrived at the street adjacent to the service company. The rain had slowed downed, just rolling showers. The roads were wet and slippery, and the lights sizzled from the rain. Aside from the occasional sound of raindrops, the whole street was quiet.

Tor glanced at the front of the service company. Three mods were standing there as if they were on guard. He stayed close to the building and slowly crept along the wall.

Suddenly, Tor heard the click of a mod in the alley to his left. He waited until the clicking sound stopped. Drawing his weapon, he lunged around the corner only to see the mod lying on the ground and a hooded figure beside him.

The man looked up and saluted Tor with the blade. Tor squinted his eyes and noticed it was another teardrop that worked for the service company. He sighed and stepped into the alley. The other teardrop was Rin, several years younger than Tor but still lean and fast as Tor remembered.

"Some man that sniffs a lot brought a bunch of mods to the company and told us we were relieved from our contracts," Rin spoke to Tor quickly. "Most of the other teardrops left back to the barracks and training grounds except Asen, Wre, and me. We are all still here." He looked around, cautious. "Something didn't seem right to us."

Tor knew them all. They were all younger and well-capable. "Where are the other two?" he asked.

Rin nodded in different directions to where the other two were. Tor acknowledged that and looked back down the street.

"Have you seen Cass?" Rin shook his head.

"Why? Is something wrong?"

"I need to find and stop her before she gets here."

"Guess you are late then," Rin said, pointing his sword to the right side. The teardrops watched Cass coming down the street. The mods in front of the service company turned to face her.

Tor looked at Rin and said, "They are trying to take her. They want her heart."

"Not today; they won't!" Rin responded and pulling out another blade from inside his cloak.

Tor turned and ran forward, charging at the mod. Out of the corner of his eye, he watched Rin run past him, blades out. And two other figures appeared on the top of the service company. Aren and Wre jumped on top of the mods at the door. Seeing his fellow teardrops covering him, Tor changed direction and went to where she was. She already had her weapon drawn.

He got closer to her and shouted, "Run, Cass! You can't be caught."

Cass looked at Tor bewildered but shrugged and turned around, running away. There were more mods behind her. And soon, more and more mods started pouring out of the buildings and alleys. Tor reached toward Cass and said, “We got to go!”

He turned back to the service company but saw the sniffing man in front of it, pointing toward him. Tor planted his feet only to hear Rin call out to him, “You get away, Tor! We will handle this trash.”

Tor nodded, and as he turned away, he saw the dance of blades and mod falling one by one as they fought against three teardrops. Cass stood watching them with amazement. She had always been with Tor in a fight but had never seen three teardrops fight together. She remembered the stories and watched what she could only perceive was ‘brutal beauty.’

Tor grabbed her arm and bellowed, “Let’s go!” He spun her around but ran straight into five mods blocking the way. Backing away slowly, he dragged her behind. “We need to go!”

Cass followed and looked back one more time. “Will the others be alright?” she asked him.

Tor didn’t reply but ran away faster, pulling her behind him. The rain was pouring again. It was difficult for them, but they still managed to get away.

“They are in the elegance of aggression now,” he grumbled. “They will be perfect.”

Cass did not know what he meant, but she knew they would be fine. The partners ran down the street together until they stopped at an alley. Panting, they noticed that no one was following them.

"Those mods have bit off more than they can handle," he said, grinning. "Cass, we must find a place to hide for a while."

"We will just go up," she said, smiling at him. He nodded and followed her as she led him to the rooftops. It was a good place to rest.

Chapter 26:
Meeting

Cass and Tor jumped from one roof to the other to the city's industrial section. She could see a group of people on the roof surrounding Ms. Sim, barking orders and directing people as usual. Frank stood nearby her, watching over her. Ben noticed the partners and walked to them. Cass could see a small knife tied to his belt.

She looked at Tor and said, "Nice gift!"

"He must keep it for himself!" Tor grumbled.

Cass smirked at him as they walked toward the group. Frank noticed them and trudged toward them. Cass felt Tor breathe in deeply and looked toward him.

"Don't worry; he is harmless. I will protect you," Cas told Tor.

"It's not that," he growled.

Cass narrowed her eyes, confused at his reaction. She glanced in the direction of where he was glaring. In the middle of the group stood a tall, hooded figure.

Mom.

Ms. Sim waved to them to come closer. Cass sighed and walked up to them, smiling. Tor stood stoic as ever. She could see him looking closely at Frank and then back at Mom. This time Mom didn't seem like she was about to tease Tor. She noticed the quick glance Mom and Tor shared between them. Tor went as stoic as ever.

What is going on? She thought. *What are they hiding from me?*

"Well, we shouldn't be standing out here in the rain," Ms. Sim interjected. "From what I understand, we will have more company coming. I guess that is an amazing heart we have right there."

She pointed to Cass's chest. Cass looked down at her chest and shrugged. Ms. Sim began walking to the building where they had dinner before and urged the others to follow. Cass walked after her and noticed that Mom glided beside her.

Mom adjusted Cass's braid and said, "You might need to fix that before you cut yourself."

Cass looked at the knife in her braid and said, "Maybe later." Mom sighed and glided away. While Cass heard Tor growl behind her.

***

Twenty minutes later, all of them were present in the makeshift dining hall. While seated, Ms. Sim directed people to bring food for their guests.

After directing, Ms. Sim turned to Cass and said, "We need to keep that heart safe." She gazed at Cass, concerned. "If Blight gets ahold of it, we don't know what will be the outcome."

"I can take her to my homeland across the mountains," Mom interjected. "The scientist that checked her before stated she should stay there for study."

One by one, everyone began voicing their opinions. Their voices overlapped the other. It frustrated Cass. She was irritated that people thought of her as a puppet that would act according to their wishes.

"Stop!" Cass shouted. Everyone stopped talking and stared at her. "I know you all are concerned for me." She looked everyone in the eye. "But it's my life, and I know what's best for me." She paused. "I am not going anywhere. It is me and not just a mechanical heart. I am not a mod, not a project, and absolutely not a child. I am Cass, and I am alive."

Silence fell across the room. Nobody moved or blinked; they just stared at her.

"Humph."

Cass heard a voice from behind her and smiled, knowing very well it was Tor, grunting in approval.

"I am not going to hide." Cass smiled at them. "And I am surely not going to leave."

She was about to speak again when suddenly the group was alerted. Turning around, the group saw three hooded figures entering the hall. Immediately, the hall went quiet, and they all glanced at each other.

Tor moved closer to Cass, alarmed but then eased off. Cass realized that it was the teardrops from the service company. Rin walked to Tor and nodded. The other two teardrops, Aren and Wre, stepped away from the group. The four teardrops huddled around each other and began talking among themselves.

They babble so quickly, Cass thought, chuckling. *Do they even talk or just growl at each other?*

After their conversation ended, Tor stepped back closer to the group and told them, "There are a large group of mods on the streets looking for us. DNA hunters are out too, but the rain is slowing them." He paused. "Rin, Wre, Aren, and I will go lead them away." He glanced at Cass. "Cass, you need to get out of here."

"I will take her with me," Mom said, stepping up.

"No!" Tor snapped quickly at her with a bellowing growl.

Mom paused and nodded, moving back from them. *Well, that's a first.* Cass thought. *What a new surprise!*

"I will go with you, Tor," Cass said, shaking her head. "What better way to stop this than to go back to Blight's factory and destroy it?"

Tor shook his head, but Mom spoke up, "I think she is right. If the factory is crippled, then Blight will have to rebuild. It will buy us some time."

"Didn't I tell you…" Tor started saying, but by that time, Mom was already moving away from him.

He glared at Mom as she glided around the table and stepped close to Cass. "Let's make a plan. With most of the mods searching in this area, we will have an advantage as there might be less at the factory."

Cass sighed and peered at Tor. The partners stared at each other for what seemed like an hour when Tor subtly nodded.

Chapter 27:
Easy Plans

An hour later, Cas cut her finger with a knife and collected it in a plastic bag. The blood was passed on to Rin, Asen, and Wre. The trio rubbed the blood on their cloaks and nodded at Tor.

It was a simple plan.

The three teardrops were going to run in circles around the street to confuse the hunters and the mods. While the enemies were distracted, Cass, Mom, and Tor, with Frank, were going to assault the factory and try to dismantle it if they could. After that, they would all gather at the train depot.

Simple! Cass thought.

Fifteen minutes later, standing behind the window, with her face hidden by a drape, Cass watched as the three teardrops walked out of the building. Giving one last look at each other, the teardrops separated and ran off in different directions.

She sighed and glanced up at the sky. The rain had stopped, and the sun was starting to set. It was the only; their only chance. Leaning away from the window, Cass realized that Tor had only told them the plan needed to be swift. And once they were done, they all had to depart to the train depot to get out of the city.

Suddenly, Cass heard a buzzing sound, and she quickly peered down the street. The three teardrops scattered on the roads with a host of scavengers in tow. The DNA hunters swiftly flew around the street and followed the three teardrops. The mods divided themselves and ran in the route the DNA hunter directed.

The plan had started. She thought, grinning.

***

Half an hour later, Tor, Cass, Mom, and Frank head swiftly to Blight's factory and central office. Cass had a concoction made of oils and what she guessed was someone's blood on her to mask the DNA. In time, the four of them reached the bridge that went to the factory.

Cass and Tor glanced at each other, nodded, and split from the other two. They ran together to an area that was partially surrounded by wood, and that led up to the factory. Meanwhile, Frank and Mom approached the front office.

Quietly, Cass and Tor leaned up against the wall; they could hear the mods walking on patrol in and around the factory. Cass modified her weapon for the blast mode. She hooked her magnetic tubes to her wristband and reached for Tor's.

Tor paused and said, "Make sure it's not in full power." He glanced around the area. "I need you to fight with me."

Cass nodded and attached the tubes from Tor's bracelet to hers. Immediately the weapon started to power up.

"Alright, here we go!" Cass commented, pointing the weapon to the wall. Tor swiftly drew his weapons and braced himself behind Cass. Leaning back slightly toward him, she felt him take deep breaths.

1....

2....

3!

At the end of the third breath, Cass pulled the trigger. The wall denigrated in front of them and blew away the mods on the other side, pushing them out of the way. The impact of the blast had Cass stumble back into Tor. And he steadied her, helping her balance.

"Are you ready?" he asked.

Cas nodded.

Tor lunged forward with Cass behind him. They waited until the dust from the blast began to settle and saw other mods and DNA hunters moving toward the opening.

"It has begun," Cass whispered.

She saw her partner grab a mod and thrash him to the ground. Without missing a beat, Tor spun with his sword and cut off another one's leg. Shaking her head, Cass focused her attention forward and changed her weapon in time to shoot a DNA hunter with a modified weapon making it roll off to the ground.

The partners swiftly moved forward, fighting mods and DNA hunters together. Tor would dodge and slash, separating the mod's arms and legs from their bodies, while Cass would shoot some mods in the chest and DNA hunters, knocking them out. They fought efficiently together, seeming to be one person.

The first shock assault was finished.

The partners hid behind a machine to take cover and look out for the enemies. Cass looked at Tor and sighed. He was covered in blood and oil that leaked out from the mods. She could've sworn she saw a slight smile on his face. It made her realize that maybe that was the joy the teardrops felt during the battle. Sensing her gaze on him, Tor glanced back at Cass and gave her a quick wink. as if answering her thought. And she smiled back at him.

***

From the front office area, Mom and Frank entered the building. The mods glanced at Mom in her hooded cloak and let her pass freely through the plaza area. The duo was supposed to destroy the lab at the science department. They walked to the elevator, entered, and with no resistance, Mom pushed the button for the 2nd floor. As soon as the elevator started going up, the alarms got activated.

Mom smirked, knowing that Cass and Tor had started the attack.

A minute later, the elevator doors opened to three mods at the entrance of the 2nd floor with weapons ready to shoot. Without giving them the upper hand, Frank sprang from the elevator at them and knocked them down. Frank also had mod parts; all four limbs were firm and fast enough to tear the other mod's limbs. Mom moved gracefully with quick steps, and with her short swords, she began dispatching the other mods. Frank was on all fours scurrying quickly in front of Mom down the corridor, moving like a scorpion. At the same time, Mom moved quickly and gracefully behind him.

They soon found the room and entered the science department. Frank grabbed another mod and started fighting with him. Meanwhile, Mom walked to the scientists that were collecting papers and tapping on screens quickly. They were busy and weren't alerted of the new intruders.

Suddenly, one of the scientists paused as Mom walked closer to him. He waited and stared at her, not comprehending the situation. While he was distracted, she smiled at him and quickly drove a sword into his stomach. The blood splattered around her while the scientist spluttered incoherent words. She lightly tapped his face as if caressing it and watched him slide down to the floor.

Huffing and rolling her eyes, Mom looked at the other two scientists in the room and calmly said, "Gather all the work and give it to me."

By then, Frank had stopped fighting with the mods and stood behind her, watching the exchange. Spooked, the scientist quickly gathered the information and handed it to her in a carry bag. Mom smiled at them; her face still hidden under her cloak.

She tilted her head toward Frank and said, "I only need one to come with me. The other can go."

Frank nodded and said, "Yes, Mom."

Mom walked away as she heard the scream of one of the scientists. Gliding to the elevator, she mumbled to herself, “Now to the factory.”

***

Before Tor and Cass could attack from behind the cover, they heard Mr. Blight's voice. Tor grunted, and Cass rolled her eyes, knowing their enemy was up to no good.

“So nice of you to bring that heart and weapon to me,” Mr. Blight commented, snickering. “Such a fool. Bringing the very same thing I plan on ripping from you.”

Tor stood up with Cass behind him and snarled, “No better day than today.”

Cas shook her head and gave Blight a wry smile. He rolled his eyes and, before the duo could react, shot his pulse wave at them. The partners quickly moved, dodging it. Frustrated, Blight grunted and directed mods and DNA hunters toward them.

Tor lunged forward at them while Cass stepped to the side to where she could get a better shot at Blight. Tor kept pushing forward, striking mods and DNA hunters alike. On the other side, Cass would shoot them as they would get closer.

Blight fired at them from the opposite side, using the hunter a shield not to get hit. This continued as Tor and Cass got closer. Mod and hunter parts were falling down in every direction. Blood, oil, and steam filled the air around Tor and Cass as they fought forward.

Tor got to Blight first, swinging his sword at him. Ducking, Blight shot the sword with his pulse wave and deflected the blade away. The teardrop recovered and struck again, but a mod stepped between them. Tor pushed through the mod while Cass shot at Blight. Unfortunately, a hunter came between them and took the fall.

Tor and Cass made several attempts to strike at Blight, but mods and hunters under his control would intervene. They still kept pressing on, trying to get to Blight. But Blight was being forced back. Before the partners could stop him, Blight came to a scaffold and glanced once last time at them.

"Shit!" Cass shouted.

"He is escaping!" Tor grumbled.

"We need to stop him, Tor!"

"Right behind you!"

Blight jumped and landed on top of a hovering bigger DNA hunter. The other hunters gathered and surrounded him. Cass blasted at several of them, and Tor ran forward, trying to reach Blight. Blight shot him with his pulse weapon, but Tor dodged it. It gave Blight the opportunity, and he began flying away with the hunters protecting him.

"Come on!" Cas shouted at Tor. "He is heading outside."

Tor and Cass started to make chase, with Blight on and surrounded by the hunter fleeing outside the building now. Cass made several more attempts to shoot the hunters but to no avail. The hunters and the mods surrounded Blight and ran with him.

The partners ran to the bridge and stopped at the bridge, watching Blight, mods, and DNA hunters flying away. Tor shook his head as Cass shot one more blast at them. Out of nowhere, Tor felt a familiar sting on his back and neck, as if something was pinching him. He quickly turned to Cas and saw her falling to the ground, darts in her back. He leaned on the bridge's railing, trying to stay on his feet.

What....? He thought. *What is going on...?*

Through his groggy vision, Tor saw Frank picking up Cass and carrying her off. A shadow appeared before him, and he realized it was Mom. She crouched down and, grinning, leaned forward, kissing his cheek.

"See you soon, Teardrop!" Mom whispered into his ear.

***

From atop his hunters, Blight watched as Mom and another figure with a scientist in tow carried Cass off. Tor was left behind on the bridge as a young boy ran toward him.

"Well played, Mom, well played!" Blight mumbled, snickering. "The heart isn't yours yet." He turned and flew away with his team, already planning his next move.

About the Author

Billy J. Rye is the son of Janice Rye, grandson of Rip and Millie Elsworth, and brother of Randy Thomas. Billy has always been part of the Marine, loves being a Dad to his kids, and wants to be a storyteller.

Made in the USA
Monee, IL
04 November 2026

67376089R00079